Mathematical Techniques in Crystallography and Materials Science

Edward Prince

Mathematical Techniques in Crystallography and Materials Science

With 28 Illustrations

Springer-Verlag
New York Heidelberg Berlin

Edward Prince, Ph.D.
National Bureau of Standards
Washington, D.C. 20234, U.S.A.

Sponsoring Editor: Philip Manor
Production: Abe Krieger

Library of Congress Cataloging in Publication Data
Prince, Edward.
Mathematical techniques in crystallography
and materials science.
Bibliography: p.
Includes index.
1. Mathematical crystallography. I. Title.
QD911.P75 548′.7 81-18443
AACR2

Printed in the United States of America.

9 8 7 6 5 4 3 2 1

ISBN 0-387-**90627-4** Springer-Verlag New York Heidelberg Berlin
ISBN 3-540-**90627-4** Springer-Verlag Berlin Heidelberg New York

Preface

In the course of 30 years as a practicing crystallographer I have frequently been faced with the necessity of finding out a little bit about some general branch of mathematics with which I was previously unfamiliar. Under these circumstances I have usually followed the common practice of seeking out some colleague who would be expected to have a thorough knowledge of the subject. I would then find myself faced either with an involved lecture in which the colleague would attempt to distill a lifetime of experience into a form that was comprehensible to a novice with a very different background, or with a book about the subject, written by a specialist, that contained far more information than I really wanted to know. I would have to separate the few kernels of useful material from a large volume of what would probably be wheat to someone else, but was chaff to me. In the course of that time I have acquired a collection of books to which I frequently refer. Most of these have a small number of thoroughly dog-eared pages, along with many that have scarcely been opened.

During the same period I have been privileged to associate and collaborate with a number of materials scientists who were not trained as crystallographers, but whose interests required them to understand particular details of some structural problem. In these cases I tried to deal with the problem at hand without requiring *them* to learn more than *they* wanted to know about crystallography. As a consequence of these experiences I have used two criteria in choosing the material to be included in this book. Either they are things that I have had to learn or look up frequently because I didn't use them enough to retain the details in ready memory, or they are things that I have frequently explained to other colleagues. It is my hope that I have covered these topics in sufficient detail to answer the day-to-day questions, or that, if the detail is not sufficient, I will have provided enough to get a person started on a fruitful search. Most active researchers will find much that they already know, but I would expect that there will be very few who will not find something that they do not know as well as they feel they should.

The level of mathematics I have assumed is that included in college courses called "Advanced Calculus," or something of that sort. It includes partial differentiation, infinite series, and improper integrals. I have not bothered with proving the convergence properties of series and integrals. The reader is asked to assume that these proofs can be found elsewhere if I use them. In the discussion of rigid-body thermal motion I have deliberately omitted any discussion of Bessel's functions. Although there are exceptions, it has been my experience that these usually appear because of a failure to replace a trigonometric function by its power series expansion at a suitable point.

My own knowledge of these topics is a result of contact with many other workers. The discussion of sublattices and superlattices is based on the work of my colleagues A. D. Mighell and A. Santoro. The discussion of constrained refinement is based on work in the course of which I have had a fruitful collaboration with L. W. Finger, and the sections on rigid-body motion owe at great deal to the work of C. K. Johnson. In the development of robust/resistant techniques I have been privileged to collaborate with W. L. Nicholson. If the discussions are correct, they owe much to the contributions of these workers, as well as to the mentors of my early years, D. Harker and W. A. Wooster. If any derivations and arguments are not correct, the blame is entirely mine.

In writing this book I have endeavored to write not a textbook but a reference book—a *vade mecum* for active research workers. There have been many times when I would have found it useful to have such a book within my reach. I hope others will also find it useful.

Bethesda, Maryland January, 1982 — Edward Prince

Contents

Chapter 1

Matrices: Definitions and Fundamental Operations

The everyday activities of the crystallographer or materials scientist deal with the real physical properties of real objects in the real, three-dimensional space in which we live. Most of these properties can be described, at least approximately, by systems of linear relationships between one set of measurable quantities, "causes," and another set of measurable quantities, "effects." We shall find that a useful shorthand for expressing these relationships is the algebra of matrices. To make use of this algebra we need to start, if only to make sure we agree on notation, with definitions of matrices and their fundamental operations.

Definition 1: A *matrix* is a two-dimensional, rectangular array of numbers. The numbers may be positive or negative, real, imaginary, or complex. In this text we denote a matrix, when referring to it as a whole, usually by a capital, boldface Roman letter, e.g., **A**. We denote the elements of a matrix by the same capital letter in regular type, and with either numbers or lowercase italic letters as subscripts, as in A_{23} or A_{ij}. The matrix has horizontal "rows" and vertical "columns." The row index is given first, so that A_{23}, above, indicates the element of **A** that is a member of the second row and of the third column. A general matrix has m rows and n columns. When necessary the matrix is written out, enclosed in parentheses:

$$\begin{pmatrix} A_{11} & A_{12} & \cdots & A_{1n} \\ A_{21} & A_{22} & \cdots & A_{2n} \\ \cdots & \cdots & \cdots & \cdots \\ A_{m1} & A_{m2} & \cdots & A_{mn} \end{pmatrix}.$$

There are certain special types of matrix it is useful to define.

Definition 2: A matrix is *square* if the number of rows is equal to the number of columns.

Definition 3: A matrix with only one row is a *row vector*.

Definition 4: A matrix with only one column is a *column vector*. We shall use column vectors frequently. They will usually be denoted by lowercase boldface letters, e.g., **x**, and their elements represented by a single subscript, as in x_i.

Definition 5: The elements of a matrix for which $i = j$ are *diagonal elements*. The elements for which $i \neq j$ are *off-diagonal elements*.

Definition 6: A matrix is *diagonal* if $A_{ij} = 0$ for all off-diagonal elements.

Definition 7: A matrix is *upper triangular* if $A_{ij} = 0$ when $i > j$.

Definition 8: A matrix is *lower triangular* if $A_{ij} = 0$ when $i < j$. Upper triangular, lower triangular, and diagonal matrices will usually also be square. A diagonal matrix is obviously also both upper triangular and lower triangular.

Definition 9: A matrix **B** is the *transpose* of **A**, designated by $\mathbf{A}^T$, if $B_{ji} = A_{ij}$. Neither **A** nor **B** need be square, but, obviously, if **A** has m rows and n columns, then **B** has n rows and m columns.

Definition 10: A *null matrix* is a matrix all of whose elements are equal to 0.

It is also useful at this point to define several explicit functions of the elements of square matrices.

Definition 11: The *trace* of the matrix **A**, designated by Tr **A**, is given by

$$\mathrm{Tr}\,\mathbf{A} = \sum_{i=1}^{n} A_{ii},$$

where the Σ denotes "the sum of all such terms as i takes all values from 1 to n." The trace is simply the sum of the diagonal elements.

Definition 12: The *determinant* of a matrix **A**, designated by $|\mathbf{A}|$ or $\operatorname{Det}\mathbf{A}$, is the sum of $n!$ (n factorial) terms of the type $(-1)^p \prod_{i=1}^{n} \mathcal{P}_p(\mathbf{A})_{ii}$. $\mathcal{P}_p(\mathbf{A})$ denotes a matrix derived from **A** by permuting columns, p being the number of times a pair of columns must be interchanged to get from **A** to $\mathcal{P}(\mathbf{A})$, whereas the Π notation denotes the product of all such factors as i takes all values from 1 to n. In the age of computers the actual value of the determinant of any matrix larger than about 3×3 rarely needs to be computed, but the function still has conceptual value, as we shall see later.

Definition 13: The ijth *minor of the determinant of* **A**, which we shall denote by $M_{ij}(\mathbf{A})$, is the determinant of the matrix derived from **A** by dropping the ith row and the jth column.

Fundamental Matrix Operations

Many of the basic operations of the arithmetic of ordinary numbers can be generalized to apply to matrices. Addition and subtraction are straightforward. The sum, **C**, of two matrices **A** and **B**, designated by $\mathbf{A} + \mathbf{B}$, is a matrix such that $C_{ij} = A_{ij} + B_{ij}$. For this definition to make sense the dimensions of all three matrices must be the same. The difference between two matrices, $\mathbf{A} - \mathbf{B}$, is correspondingly the matrix **C** such that $C_{ij} = A_{ij} - B_{ij}$. Addition and subtraction obey the associative law—$(\mathbf{A} + \mathbf{B}) + \mathbf{C} = \mathbf{A} + (\mathbf{B} + \mathbf{C})$—and also the commutative law—$\mathbf{A} + \mathbf{B} = \mathbf{B} + \mathbf{A}$.

Multiplication of matrices comes in two forms. A matrix may be multiplied by a scalar number. The product of a scalar a with a matrix **A** is the matrix **B** such that $B_{ij} = aA_{ij}$. The dimensions of **A** and **B** are equal. The product **C** of two matrices, **A** and **B**, is the matrix such that

$$C_{ik} = \sum_{j=1}^{n} A_{ij} B_{jk}.$$

n is the number of columns of **A** and the number of rows of **B**, which must be equal.[1] When defined in this way, matrix multiplication obeys the associative law: $(\mathbf{AB})\mathbf{C} = \mathbf{A}(\mathbf{BC})$. It does not obey the commutative law: $\mathbf{AB} \neq \mathbf{BA}$. To form the transpose of a product it is necessary to interchange the multiplicands: $(\mathbf{AB})^T = \mathbf{B}^T\mathbf{A}^T$. To see that this is true it is necessary only to write out the expression for an element of the product. The

[1] In many publications a convention is used in which the sum is implied if the letters used for the subscripts are repeated. In other words, $A_{ij}B_{jk}$ is taken to mean $\sum_{j=1}^{n} A_{ij}B_{jk}$. I have always found the use of this "summing convention" confusing and often frustrating when I was trying to follow a derivation or proof. For this reason the convention will not be used any further in this book.

determinant of a product matrix, $|\mathbf{AB}|$, is equal to the product of the determinants of the factor matrices: $|\mathbf{AB}| = |\mathbf{A}|\,|\mathbf{B}|$.

There is no analog to division for general matrices. Division can, however, be thought of as multiplication by a reciprocal, and there is an analog to forming a reciprocal for many square matrices. Before discussing this operation we shall define another special matrix, the *identity matrix*.

Definition 14: The $n \times n$ *identity matrix*, designated by $\mathbf{I}_n$, or simply by $\mathbf{I}$ if the dimensions are not important in a particular discussion, is the square matrix such that $I_{ij} = 1$ if $i = j$ and $I_{ij} = 0$ if $i \neq j$. In other words, it is a matrix with 1 value on the main diagonal and 0 values elsewhere. (A function that is equal to 1 if $i = j$ and 0 if $i \neq j$ is often referred to as the *Kronecker delta function* and is denoted by δ_{ij}.) It can readily be determined that $\mathbf{AI} = \mathbf{A}$ and $\mathbf{IA} = \mathbf{A}$. In each case $\mathbf{I}$ is assumed to have the proper dimensions for the multiplication to be performed.

Definition 15: The *inverse* of a matrix $\mathbf{A}$, denoted by $\mathbf{A}^{-1}$, is a matrix such that $\mathbf{AA}^{-1} = \mathbf{A}^{-1}\mathbf{A} = \mathbf{I}$. The inverse of the matrix can be found by solving the system of n^2 simultaneous equations of the form

$$\sum_{j=1}^{n} A_{ij}A_{jk}^{-1} = \delta_{ik}.$$

We shall not consider the straightforward but tedious algebra, but shall simply state the result.

$$A_{ij}^{-1} = (-1)^{i+j}M_{ji}(\mathbf{A})/|\mathbf{A}|.$$

Since division by 0 is undefined, this operation cannot be performed if $|\mathbf{A}| = 0$. A matrix whose determinant is equal to zero is said to be *singular*. A singular matrix has no inverse.

The inverse of a 3×3 matrix can be easily written out explicitly. Let $\Delta = |\mathbf{A}| = A_{11}(A_{22}A_{33} - A_{23}A_{32}) + A_{21}(A_{32}A_{13} - A_{12}A_{33}) + A_{31}(A_{12}A_{23} - A_{13}A_{22})$. Then

$$\mathbf{A}^{-1} = \begin{bmatrix} (A_{22}A_{33} - A_{23}A_{32})/\Delta & -(A_{12}A_{33} - A_{13}A_{32})/\Delta & (A_{12}A_{23} - A_{13}A_{22})/\Delta \\ -(A_{21}A_{33} - A_{31}A_{23})/\Delta & (A_{11}A_{33} - A_{13}A_{31})/\Delta & -(A_{11}A_{23} - A_{13}A_{21})/\Delta \\ (A_{21}A_{32} - A_{31}A_{22})/\Delta & -(A_{11}A_{32} - A_{12}A_{31})/\Delta & (A_{11}A_{22} - A_{12}A_{21})/\Delta \end{bmatrix}.$$

Although the inverse is defined only for a square matrix, there are situations in matrix algebra where a division would normally be called for in simple algebra that have analogs involving rectangular matrices. In these cases we often find that the matrix $\mathbf{B} = (\mathbf{A}^T\mathbf{A})^{-1}\mathbf{A}^T$ appears. This form is, in a sense, an analog to a reciprocal for a general matrix. Obviously $(\mathbf{A}^T\mathbf{A})$ must not be singular.

Because a matrix, as we have defined the term, is simply a rectangular array of numbers, matrices can be formed by taking rectangular subdivisions of larger matrices, and larger ones can be formed by constructing rectangular arrays of smaller ones. The process of making two or more smaller matrices out of a large one is called *partition*. The corresponding process of making a large matrix out of two or more smaller ones is called *augmentation*.

Linear Algebra

A large fraction of the mathematical operations in materials science involves the solution of systems of linear equations. Thus

$$
\begin{aligned}
A_{11}x_1 + A_{12}x_2 + A_{13}x_3 + \cdots + A_{1n}x_n &= b_1, \\
A_{21}x_1 + A_{22}x_2 + A_{23}x_3 + \cdots + A_{2n}x_n &= b_2, \\
A_{31}x_1 + A_{23}x_2 + A_{33}x_3 + \cdots + A_{3n}x_n &= b_3, \\
\cdots \quad \cdots \quad \cdots \quad \cdots \quad \cdots \quad &\cdots, \\
A_{n1}x_1 + A_{n2}x_2 + A_{n3}x_3 + \cdots + A_{nn}x_n &= b_n.
\end{aligned}
$$

It will be immediately recognized that this system of equations can be represented in a compact shorthand by the single matrix equation

$$\mathbf{Ax} = \mathbf{b},$$

where $\mathbf{A}$ is a matrix whose elements are A_{11}, A_{12}, etc., and $\mathbf{x}$ and $\mathbf{b}$ are column vectors. Just as in basic algebra, if the same operation is performed on both sides of an equation, the equation remains an equation. If the matrix $\mathbf{A}$ is not singular, it has an inverse, $\mathbf{A}^{-1}$. Thus

$$\mathbf{A}^{-1}\mathbf{Ax} = \mathbf{A}^{-1}\mathbf{b},$$

or

$$\mathbf{x} = \mathbf{A}^{-1}\mathbf{b}.$$

Therefore, the equation can be solved for the unknown vector **x** by finding the inverse of the matrix **A** and multiplying the known vector **b** by this inverse. In practice this is almost never done, except for cases of very small systems of equations, because expanding a determinant, which involves a sum of $n!$ terms each of which is the product of n factors, quickly adds up to an enormous number of excessively time-consuming arithmetic operations, even with the use of a high-speed computer. There are many algorithms for solving the equations—because computers have finite precision, different algorithms work best in different particular circumstances—but they all conceptually follow the same steps.

1. Find a lower triangular matrix, **L**, and an upper triangular matrix, **U**, such that **LU** = **A**. Two specific procedures for performing this step, called decomposition, are described on the following pages.
2. Invert **L**. This is relatively easy to do by following the procedure:

$$L_{11}L_{11}^{-1} = 1; \text{ therefore } L_{11}^{-1} = 1/L_{11};$$

$$L_{21}L_{11}^{-1} + L_{22}L_{21}^{-1} = 0; \text{ so } L_{21}^{-1} = -L_{21}/(L_{11}L_{22});$$

and so forth. By this method all nonzero elements of $\mathbf{L}^{-1}$ (which is also lower triangular) can be evaluated from expressions involving only the elements of **L** and previously evaluated elements of $\mathbf{L}^{-1}$.
3. Multiply both sides of the equation by $\mathbf{L}^{-1}$, giving the equation

$$\mathbf{Ux} = \mathbf{L}^{-1}\mathbf{b},$$

or

$$\mathbf{Ux} = \mathbf{c}.$$

4. Solve these equations by back-substitution, i.e.,

$$x_n = c_n/U_{nn},$$

$$x_{n-1} = (c_{n-1} - U_{n-1,n}x_n)/U_{n-1,n-1}, \text{ etc.}$$

The actual inverse of **A** is seldom of interest, but if required, as in some statistical analyses, it can be determined by first finding $\mathbf{U}^{-1}$ and then forming the product $\mathbf{U}^{-1}\mathbf{L}^{-1} = \mathbf{A}^{-1}$.

A specific example of a procedure for solving systems of linear equations, which can be used on general nonsingular matrices, is called *Gaussian elimination*. It is performed as follows:

1. Multiply both $\mathbf{A}$ and $\mathbf{b}$ by the lower triangular matrix $\mathbf{L}'$, defined by

$$\mathbf{L}' = \begin{bmatrix} 1 & 0 & 0 & \dots & 0 \\ -A_{21}/A_{11} & 1 & 0 & \dots & 0 \\ -A_{31}/A_{11} & 0 & 1 & \dots & 0 \\ \cdot & \cdot & \cdot & \cdot & \cdot \\ -A_{n1}/A_{11} & 0 & 0 & \dots & 1 \end{bmatrix}.$$

Denote the product $\mathbf{L}'\mathbf{A}$ by $\mathbf{A}'$. Then

$$A'_{11} = A_{11};$$

$$A'_{i1} = -(A_{i1}A_{11})/A_{11} + A_{i1} = 0, \text{ for } i > 1.$$

The matrix $\mathbf{A}'$ therefore has all of the elements in the first column equal to zero except the first one, which is unchanged.

2. Multiply both $\mathbf{A}'$ and $\mathbf{b}'$ $(= \mathbf{L}'\mathbf{b})$ by the matrix $\mathbf{L}''$, defined by

$$\mathbf{L}'' = \begin{bmatrix} 1 & 0 & 0 & \dots & 0 \\ 0 & 1 & 0 & \dots & 0 \\ 0 & -A'_{32}/A'_{22} & 1 & \dots & 0 \\ \cdot & \cdot & \cdot & \cdot & \cdot \\ 0 & -A'_{n2}/A'_{22} & 0 & \dots & 1 \end{bmatrix}.$$

Denoting the product of $\mathbf{L}''\mathbf{A}$, as before, by $\mathbf{A}''$, we have

$$A''_{22} = A'_{22}$$

$$A''_{i2} = -(A'_{i2}A'_{22})/A'_{22} + A'_{i2} = 0 \text{ for } i > 2.$$

Thus $\mathbf{A}''$ has zeros below the diagonal in the first two columns.

3. Repeat this process $n - 3$ more times, each time putting zeros below the diagonal in one more column. The product $\mathbf{L}^{(n-1)} \dots \mathbf{L}''\mathbf{L}'\mathbf{A}$ will then be upper triangular, and we can equate it to $\mathbf{U}$. Further, the product $L^{(n-1)} \dots \mathbf{L}''\mathbf{L}'$ is $\mathbf{L}^{-1}$, and, since we have been multiplying $\mathbf{b}$, $\mathbf{b}'$, $\mathbf{b}''$, etc. in turn as we went along, we now have all of the coefficients of our system of equations $\mathbf{Ux} = \mathbf{c}$, and we can proceed with the back-substitution to find $\mathbf{x}$.

In the lower triangular matrix $\mathbf{L}^{-1}$ all diagonal elements are equal to 1, and therefore its determinant is equal to 1. Since $\mathbf{U} = \mathbf{L}^{-1}\mathbf{A}$, $|\mathbf{U}| = |\mathbf{A}|$, but $|\mathbf{U}|$ is the product of its diagonal elements. This is the most efficient way to evaluate the determinant of a general matrix, if it is of interest.

In Chapter 6, in an example of least squares fitting, we shall need to solve the system of equations

$$\begin{aligned} 10x_1 + \quad 50x_2 + \quad 375x_3 &= \ 15.33, \\ 50x_1 + \ 375x_2 + \ 3125x_3 &= \ 99.75, \\ 375x_1 + 3125x_2 + 27344x_3 &= 803.27, \end{aligned}$$

or $\mathbf{Ax} = \mathbf{v}$, *where*

$$\mathbf{A} = \begin{bmatrix} 10.0 & 50.0 & 375.0 \\ 50.0 & 375.0 & 3125.0 \\ 375.0 & 3125.0 & 27344.0 \end{bmatrix},$$

and

$$\mathbf{v} = \begin{bmatrix} 15.33 \\ 99.75 \\ 803.27 \end{bmatrix}.$$

These may be solved by Gaussian elimination as follows:

$$\mathbf{L}' = \begin{bmatrix} 1.0 & 0.0 & 0.0 \\ -5.0 & 1.0 & 0.0 \\ -37.5 & 0.0 & 1.0 \end{bmatrix},$$

so that

$$\mathbf{L}'\mathbf{A} = \begin{bmatrix} 10.0 & 50.0 & 375.0 \\ 0.0 & 125.0 & 1250.0 \\ 0.0 & 1250.0 & 13281.0 \end{bmatrix} = \mathbf{A}'$$

and

$$\mathbf{L}'\mathbf{v} = \begin{bmatrix} 15.33 \\ 23.10 \\ 228.36 \end{bmatrix} = \mathbf{v}'.$$

$$\mathbf{L}'' = \begin{bmatrix} 1.0 & 0.0 & 0.0 \\ 0.0 & 1.0 & 0.0 \\ 0.0 & -10.0 & 1.0 \end{bmatrix},$$

so that

$$\mathbf{L}''\mathbf{A}' = \begin{bmatrix} 10.0 & 50.0 & 375.0 \\ 0.0 & 125.0 & 1250.0 \\ 0.0 & 0.0 & 781.0 \end{bmatrix},$$

and

$$\mathbf{L}''\mathbf{v}' = \begin{bmatrix} 15.33 \\ 23.10 \\ -2.63 \end{bmatrix}.$$

Solving by back-substitution,

$$781x_3 = -2.63, \textit{ or } x_3 = -0.00337;$$

$$125x_2 = 23.10 + 4.21, \textit{ or } x_2 = 0.2185;$$

$$10x_1 = 15.33 - 10.92 + 1.26, \textit{ or } x_1 = 0.567.$$

Eigenvalues

We shall see later that we are often interested in finding a vector, $\mathbf{x}$, which, when multiplied by a matrix, $\mathbf{A}$, gives a result that is a scalar multiple of $\mathbf{x}$. In other words, $\mathbf{Ax} = \lambda\mathbf{x}$. This matrix equation is shorthand for the system of linear equations

$$A_{11}x_1 + A_{12}x_2 + \cdots + A_{1n}x_n = \lambda x_1,$$

$$A_{21}x_1 + A_{22}x_2 + \cdots + A_{2n}x_n = \lambda x_2,$$

$$\cdots \quad \cdots \quad \cdots$$

$$A_{n1}x_1 + A_{n2}x_2 + \cdots + A_{nn}x_n = \lambda x_n.$$

These equations can be rewritten

$$(A_{11} - \lambda)x_1 + A_{12}x_2 + \cdots + A_{1n}x_n = 0,$$

$$A_{21}x_1 + (A_{22} - \lambda)x_2 + \cdots + A_{2n}x_n = 0,$$

$$\cdots \quad \cdots \quad \cdots$$

$$A_{n1}x_1 + A_{n2}x_2 + \cdots (A_{nn} - \lambda)x_n = 0,$$

or, as a matrix equation, $(\mathbf{A} - \lambda\mathbf{I})\mathbf{x} = \mathbf{0}$, where $\mathbf{0}$ denotes a null vector with the dimensions of $\mathbf{x}$. We have already seen that if the matrix $(\mathbf{A} - \lambda\mathbf{I})$ is nonsingular, it has an inverse, and the solution is $\mathbf{x} = (\mathbf{A} - \lambda\mathbf{I})^{-1}\mathbf{0} = \mathbf{0}$, so that for a general value of λ, the only solution is the trivial one that $\mathbf{x}$ is a null vector. If the matrix is singular, however, the system of equations is underdetermined and has an infinite number of solutions. This fact leads to an equation, commonly known as the secular equation, $|\mathbf{A} - \lambda\mathbf{I}| = 0$. The expansion of the determinant gives a polynomial of degree n whose leading term is $(-\lambda)^n$ and whose constant term is $|\mathbf{A}|$. The polynomial will have n roots that are known as the *eigenvalues* of the matrix **A**. The word eigenvalue is a coined English word to serve as a translation of the otherwise untranslatable German word *Eigenwert*. In some older literature the term "characteristic value" sometimes appears, but in the recent literature eigenvalue is used exclusively.

In physical science we frequently observe matrices that have element A_{ij} equal to A_{ji}^*, where the asterisk denotes complex conjugate. This leads us to introduce two more definitions.

Definition 16: The *adjoint* or *conjugate transpose* of a matrix, **A**, denoted by $\mathbf{A}^\dagger$, is a matrix such that $A_{ij}^\dagger = A_{ji}^*$.

Definition 17: A matrix that is equal to its adjoint is *hermitian*. If all of the elements of the hermitian matrix are real, the matrix is *symmetric*. A hermitian matrix has the useful property that it can be proved that all of its eigenvalues must be real. Consequently, all of the eigenvalues of a symmetric matrix are real.

Before we discuss the method of finding the eigenvalues of a matrix, we need additional definitions of special matrices.

Definition 18: A *unitary matrix* is a matrix, **A**, such that $\mathbf{A}^{-1} = \mathbf{A}^\dagger$. We can see that, because $|\mathbf{A}^{-1}| = 1/|\mathbf{A}|$, and $|\mathbf{A}| = |\mathbf{A}^\dagger|$, $|\mathbf{A}| = \pm 1$ if **A** is unitary.

Definition 19: An *orthogonal* matrix is a unitary matrix all of whose elements are real. We can see immediately that $\mathbf{A}^{-1} = \mathbf{A}^T$ if **A** is orthogonal. Since $\mathbf{A}\mathbf{A}^T = \mathbf{I}$, it follows that

$$\sum_{k=1}^{n} A_{ik}A_{jk} = \delta_{ij},$$

or, in other words, the sum of the squares of the elements in any row (column) is equal to one, whereas the sum of the products of elements from one row (column) and the corresponding elements of another row (column) is equal to zero.

Definition 20: A *similarity transformation* of a matrix, **A**, is a matrix product of the form $\mathbf{T}^{-1}\mathbf{AT}$. Clearly, if the transformation matrix, **T**, is unitary, then the product $\mathbf{T}^{\dagger}\mathbf{AT}$ is a similarity transformation; correspondingly, if **T** is orthogonal, then $\mathbf{T}^{T}\mathbf{AT}$ is a similarity transformation. A unitary (orthogonal), similarity transformation applied to a hermitian (symmetric) matrix leaves the matrix hermitian (symmetric).

We have seen that the determinant of the product of two matrices is equal to the product of the determinants. It follows that the similarity transformation of the matrix $(\mathbf{A} - \lambda\mathbf{I})$ will give a new matrix with the same determinant. Since equating the determinant of this matrix to zero gives us the secular equation whose roots are the eigenvalues, we can see immediately that a similarity transformation leaves the eigenvalues unchanged. If we can find, therefore, a matrix, **T**, such that $\mathbf{T}^{-1}\mathbf{AT}$ is diagonal, the diagonal elements of the product matrix are its eigenvalues.

In practice, the methods for determining eigenvalues depend on the size of the matrix. Because we live in a space that has three dimensions, many problems involve the eigenvalues of a 3×3 matrix. In this case the secular equation is cubic, and methods exist for solving it directly. (The roots of a cubic equation and the eigenvalues of 3×3 matrices are discussed in Appendix A.) For larger matrices the methods involve iterative procedures for finding a unitary, similarity transformation that yields a diagonal matrix. The rows of the transformation matrix are vectors that satisfy the equation by which eigenvalues are defined, and, for this reason, are called *eigenvectors*.

A similarity transformation also leaves the trace of a matrix unchanged. That this is true may be seen by noting that

$$A'_{ii} = \sum_{k=1}^{n} \sum_{l=1}^{n} T_{il}^{-1} A_{lk} T_{ki},$$

so that

$$\mathrm{Tr}(\mathbf{A}') = \sum_{i=1}^{n} A'_{ii} = \sum_{i=1}^{n} \sum_{k=1}^{n} \sum_{l=1}^{n} T_{il}^{-1} A_{lk} T_{ki}.$$

Since we are summing a finite number of terms, we can rearrange the order of summation, giving

$$\mathrm{Tr}(\mathbf{A}') = \sum_{k=1}^{n} \sum_{l=1}^{n} A_{lk} \sum_{i=1}^{n} T_{il}^{-1} T_{ki},$$

but

$$\sum_{i=1}^{n} T_{il}^{-1} T_{ki} = \delta_{lk}.$$

Therefore

$$\mathrm{Tr}(\mathbf{A}') = \sum_{k=1}^{n} A_{kk} = \mathrm{Tr}(\mathbf{A}).$$

Definition 21: A symmetric matrix, $\mathbf{A}$, for which the product $\mathbf{x}^T\mathbf{A}\mathbf{x}$ is positive for any vector $\mathbf{x}$ is said to be *positive definite*. Since the vector $\mathbf{x}$ can be, in particular, any one of its eigenvectors, for which the product reduces to $\lambda\sum_{i=1}^{n} x_i^2$, it follows that all eigenvalues of a positive definite matrix are positive.

If a matrix is positive definite, it can be shown that there exists an upper triangular matrix, $\mathbf{U}$, such that $\mathbf{U}^T\mathbf{U} = \mathbf{A}$. The matrix $\mathbf{U}$ is called the *Cholesky decomposition*, or, sometimes, the *upper triangular square root* of $\mathbf{A}$. For a large matrix the procedure for determining the elements of $\mathbf{U}$ involves many fewer operations than the procedure of Gaussian elimination described earlier, so that it can be used in a more efficient procedure for matrix inversion. To see how the procedure works we can expand the product $\mathbf{U}^T\mathbf{U}$.

$$A_{11} = U_{11}^2; \quad \text{so } U_{11} = A_{11}^{1/2}.$$

$$A_{1i} = U_{11}U_{1i}; \quad \text{so } U_{1i} = A_{1i}/U_{11} \quad \text{for } i > 1.$$

$$A_{22} = U_{12}^2 + U_{22}^2, \quad \text{so } U_{22} = \left(A_{22} - U_{12}^2\right)^{1/2}.$$

$$A_{2i} = U_{12}U_{1i} + U_{22}U_{2i}, \quad \text{so } U_{2i} = (A_{21} - U_{12}U_{1i})/U_{22} \quad \text{for } i > 2.$$

In general

$$U_{ii} = \left(A_{ii} - \sum_{j=1}^{i-1} U_{ji}^2\right)^{1/2}$$

$$U_{ij} = \left(A_{ij} - \sum_{k=1}^{j-1} U_{ki}U_{kj}\right)/U_{ii}, \text{ for } j > i.$$

The proof that such a procedure is always possible for a positive definite matrix involves showing that all the quantities whose square root must be determined to compute the diagonal elements of $\mathbf{U}$ are positive, so that the square root is real. All elements of $\mathbf{U}$ are computed in sequence by expressions that involve only the corresponding elements of $\mathbf{A}$ and elements of $\mathbf{U}$ that have been computed previously.

The matrix for the system of equations in the box on page 8 is positive definite. The equations may therefore be solved by Cholesky decomposition as follows:

$$\mathbf{A} = \mathbf{U}^T\mathbf{U} = \begin{bmatrix} 10.0 & 50.0 & 375.0 \\ 50.0 & 375.0 & 3125.0 \\ 375.0 & 3125.0 & 27344.0 \end{bmatrix}.$$

$$\mathbf{U} = \begin{bmatrix} 3.1622 & 15.8114 & 118.5854 \\ 0.0000 & 11.1803 & 111.8035 \\ 0.0000 & 0.0000 & 27.9509 \end{bmatrix}.$$

$$\mathbf{L}^{-1} = (\mathbf{U}^{-1})^T = \begin{bmatrix} 0.316228 & 0.000000 & 0.000000 \\ -0.447214 & 0.089443 & 0.000000 \\ 0.447212 & -0.357770 & 0.035777 \end{bmatrix},$$

and

$$\mathbf{L}^{-1}\mathbf{v} = \begin{bmatrix} 4.8481 \\ 2.0660 \\ -0.0921 \end{bmatrix}.$$

Solving by back-substitution gives

$$27.9509x_3 = -0.0942, \text{ or } x_3 = -0.00337;$$

$$11.1803x_2 = 2.0660 + 0.3768, \text{ or } x_2 = 0.2185;$$

$$3.1622x_1 = 4.8481 - 3.4548 + 0.3996, \text{ or } x_1 = 0.5670,$$

which is identical to the result obtained by Gaussian elimination. From $\mathbf{U}^{-1}\mathbf{L}^{-1}$ *we can get* $\mathbf{A}^{-1}$, *which is*

$$\mathbf{A}^{-1} = \begin{bmatrix} 0.50000 & -0.20000 & 0.01600 \\ -0.20000 & 0.13600 & -0.01280 \\ 0.01600 & -0.012800 & 0.00128 \end{bmatrix}.$$

Linear Transformations

Much of materials science involves the measurement of things in a way that requires the use of some frame of reference, a set of coordinate axes. Very often we are working in three-dimensional real space, and we are dealing with things that work out most easily if the frame of reference is a cartesian coordinate system based on three mutually perpendicular axes along which distances are measured in some common unit. Such a system is referred to as an *orthonormal coordinate system*. At other times it is convenient to use a coordinate system in which the axes are not mutually perpendicular and/or do not have a common unit of measure. The locations of the equilibrium positions of atoms with respect to the lattice points of a noncubic space lattice is an example of this: the basis vectors of the lattice provide natural units of measure in a system that is not orthonormal. If we wish to describe displacements of atoms from their equilibrium positions, it is often convenient to use each equilibrium position as a separate origin, with three components of displacement from the origin, and then use, as the basic measure, some linear combination of those components, called a *normal mode*. In this case we are working in a virtual space of many more than three dimensions, but each state of the system is described by a sequence of numbers, a vector. In all of these cases it is necessary to make extensive use of relationships between coordinates referred to one system of axes and coordinates referred to another system of axes. These relationships are usually linear, at least to an adequate degree of approximation, and they can be easily dealt with by using matrix algebra. Such relationships are known as *linear transformations*.

Rotation of Axes

One of the simplest forms of linear transformation is a rotation of axes about a fixed origin. Figure 1.1 shows two orthonormal coordinate systems in two dimensions, with a common origin and an angle α included between axis x_1 and axis x_1'. The coordinates of point P are a_1 and a_2 in reference to axes x_1 and x_2, and b_1 and b_2 in reference to axes x_1' and x_2'. We can see, from elementary trigonometry, that b_1 and b_2 are related to a_1 and a_2 by the relationships

$$b_1 = a_1 \cos\alpha + a_2 \sin\alpha,$$

$$b_2 = -a_1 \sin\alpha + a_2 \cos\alpha.$$

We can see, therefore, that we can generate the vector

$$\mathbf{b} = \begin{pmatrix} b_1 \\ b_2 \end{pmatrix}$$

by multiplying the vector

$$\mathbf{a} = \begin{pmatrix} a_1 \\ a_2 \end{pmatrix}$$

by the matrix

$$\mathbf{R} = \begin{pmatrix} \cos\alpha & \sin\alpha \\ -\sin\alpha & \cos\alpha \end{pmatrix}.$$

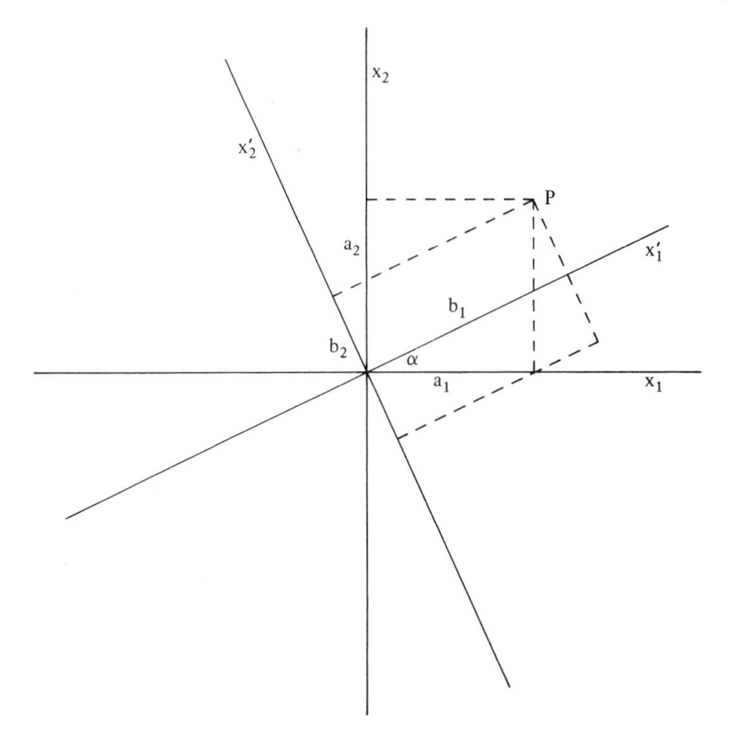

Fig. 1.1. Transformation of axes in two dimensions.

We can also see that the inverse relationships are

$$a_1 = b_1 \cos\alpha - b_2 \sin\alpha,$$

$$a_2 = b_1 \sin\alpha + b_2 \cos\alpha,$$

so that the inverse matrix is

$$\mathbf{R}^{-1} = \begin{pmatrix} \cos\alpha & -\sin\alpha \\ \sin\alpha & \cos a \end{pmatrix} = \mathbf{R}^T.$$

The matrix **R** is therefore orthogonal.

For two successive rotations of axes, the first through an angle α and the second through an angle β, the matrix is

$$\mathbf{R} = \begin{pmatrix} \cos(\alpha+\beta) & \sin(\alpha+\beta) \\ -\sin(\alpha+\beta) & \cos(\alpha+\beta) \end{pmatrix},$$

or

$$\mathbf{R} = \begin{pmatrix} \cos\alpha\cos\beta - \sin\alpha\sin\beta & \sin\alpha\cos\beta + \cos\alpha\sin\beta \\ -\sin\alpha\cos\beta - \cos\alpha\sin\beta & \cos\alpha\cos\beta - \sin\alpha\sin\beta \end{pmatrix},$$

which can be seen to be the matrix product of two matrices, $\mathbf{R}_\alpha \mathbf{R}_\beta$, where

$$\mathbf{R}_\alpha = \begin{pmatrix} \cos\alpha & \sin\alpha \\ -\sin\alpha & \cos\alpha \end{pmatrix},$$

and

$$\mathbf{R}_\beta = \begin{pmatrix} \cos\beta & \sin\beta \\ -\sin\beta & \cos\beta \end{pmatrix}.$$

In two dimensions the transformation matrix corresponding to a rotation within the plane has dimensions 2×2, with four elements, but they are all specified by a single parameter, the angle α. In three dimensions the corresponding operation is described by 3×3 orthogonal matrix whose elements are the direction cosines of the new axes with respect to the old ones. The matrix has nine elements, but there are six relationships among them, corresponding to the set of equations

$$\sum_{k=1}^{3} A_{ik} A_{jk} = \delta_{ij},$$

leaving only three parameters to be specified. There are various ways to specify these three parameters, but one of the most common and most convenient is by means of three angles, known as *Eulerian angles*, after the mathematician Leonhard Euler.

By referring to Figure 1.2 we can see how these angles are defined. First, the coordinates are rotated about the x_3 axis by an angle ω until the new x_3 axis, which will be designated x'''_3, lies in the plane defined by x_3 and x'_1. This transformation is performed by the orthogonal matrix

$$\mathbf{\Omega} = \begin{bmatrix} \cos\omega & \sin\omega & 0 \\ -\sin\omega & \cos\omega & 0 \\ 0 & 0 & 1 \end{bmatrix}.$$

Next, rotate the coordinates about the x'_2 axis (which is now perpendicular to x_3 and x'''_3) through an angle χ until the x'_3 and x''_3 axes coincide. The matrix for this transformation is

$$\mathbf{X} = \begin{bmatrix} \cos\chi & 0 & -\sin\chi \\ 0 & 1 & 0 \\ \sin\chi & 0 & \cos\chi \end{bmatrix}.$$

Finally, rotate the coordinates about $x''_3 (= x'''_3)$ by an angle ϕ until the x''_1 and x''_2 axes coincide with x'''_1 and x'''_2. This transformation is

$$\mathbf{\Phi} = \begin{bmatrix} \cos\phi & \sin\phi & 0 \\ -\sin\phi & \cos\phi & 0 \\ 0 & 0 & 1 \end{bmatrix}.$$

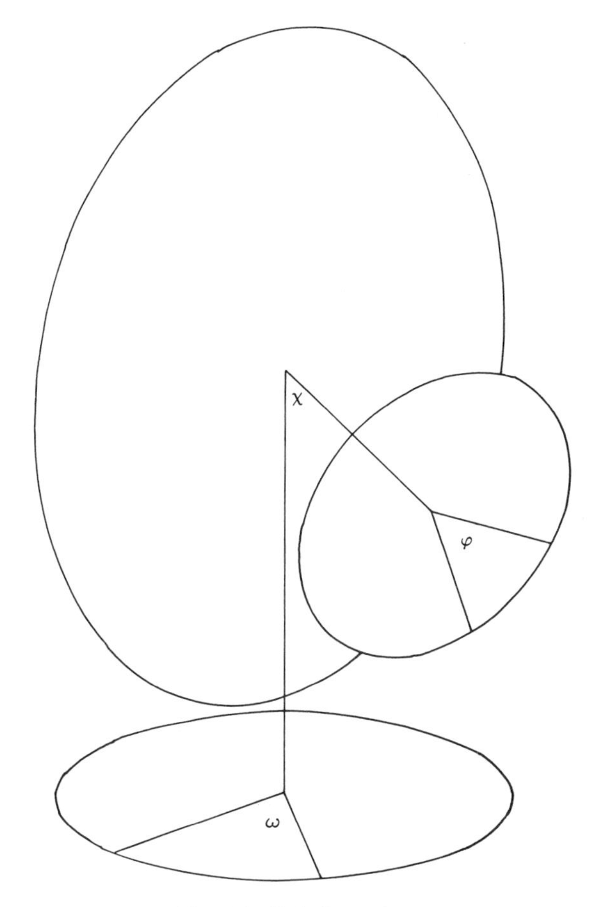

Fig. 1.2. Euler's angles.

The overall transformation is given by

$$\mathbf{R} = \mathbf{\Phi X \Omega} = \begin{pmatrix} \cos\phi\cos\chi\cos\omega - \sin\phi\sin\omega & \cos\phi\cos\chi\sin\omega + \sin\phi\cos\omega & -\cos\phi\sin\chi \\ -\sin\phi\cos\chi\cos\omega - \cos\phi\sin\omega & -\sin\phi\cos\chi\sin\omega + \cos\phi\cos\omega & \sin\phi\sin\chi \\ \sin\chi\cos\omega & \sin\chi\sin\omega & \cos\chi \end{pmatrix}.$$

The secular equation for this 3×3 matrix is cubic, so that it must have at least one real eigenvalue. If we transform the matrix into a coordinate system in which the corresponding eigenvector is one of the axes, the matrix reduces to a rotation around that axis. Thus any general transformation of coordinate axes is equivalent to a rotation about some axis passing through the origin. This result is known as *Euler's theorem*.

The Metric Tensor

Another important linear transformation is the conversion of vectors defined in a crystal space, as fractions of the basis vectors, **a**, **b**, and **c** of a crystal lattice, to an orthonormal coordinate system, with axes x, y, and z, in which interatomic distances and bond angles can be conveniently calculated. We shall make extensive use of a "standard" orthonormal system related to the crystal lattice in the following way: The x axis is parallel to **a**. The y axis lies in the $\mathbf{a}-\mathbf{b}$ plane, perpendicular to x. The z axis is perpendicular to x and y, and forms a right-handed coordinate system with them. Referring to Figure 1.3, we can see the following relationships: the projection of **a** on x is $|\mathbf{a}|$; the projection of **b** on x is $|\mathbf{b}|\cos\gamma$ and on y is $|\mathbf{b}|\sin\gamma$; the projections of **c** on x and y are $|\mathbf{c}|\cos\beta$ and $|\mathbf{c}|(\cos\alpha - \cos\beta\cos\gamma)/\sin\gamma$. Because the trigonometric coefficients of $|\mathbf{c}|$ are the direction cosines, and because the sum of squares of the three direction cosines must be one, the projection of **c** on z is $|\mathbf{c}|[1-(\cos^2\alpha + \cos^2\beta - 2\cos\alpha\cos\beta\cos\gamma)]^{1/2}/\sin\gamma$. These relationships can be used to define an upper triangular matrix

$$\mathbf{A} = \begin{bmatrix} a & b\cos\gamma & c\cos\beta \\ 0 & b\sin\gamma & c(\cos\alpha - \cos\beta\cos\gamma)/\sin\gamma \\ 0 & 0 & c\left[1-(\cos^2\alpha + \cos^2\beta - 2\cos\alpha\cos\beta\cos\gamma)\right]^{1/2}/\sin\gamma \end{bmatrix}.$$

The magnitude of a vector, **d**, is $(\mathbf{d}^T\mathbf{d})^{1/2}$, and, if its components are expressed in crystalline, fractional units, its magnitude in real space units is

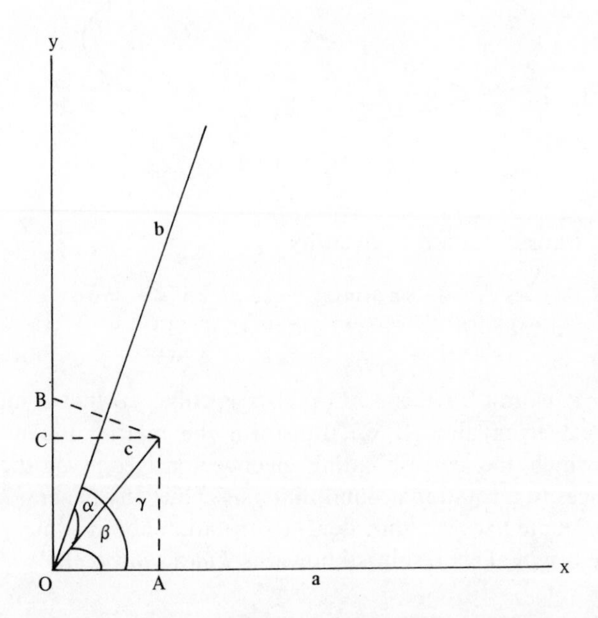

Fig. 1.3. Transformation from crystal axes to an orthonormal coordinate system. View is down the z axis. **b** is in the $x-y$ plane, and **c** comes out of the paper. $\overline{OB} = |\mathbf{c}|\cos\alpha/\sin\gamma$; $\overline{OA} = |\mathbf{c}|\cos\beta$; $\overline{BC} = \overline{OA}\,\mathrm{ctn}\,\gamma = |\mathbf{c}|\cos\beta\cos\gamma/\sin\gamma$.

given by $|\mathbf{d}| = (\mathbf{d}^T\mathbf{A}^T\mathbf{A}\mathbf{d})^{1/2}$, or $|\mathbf{d}| = (\mathbf{d}^T\mathbf{G}\mathbf{d})^{1/2}$, where

$$\mathbf{G} = \mathbf{A}^T\mathbf{A} = \begin{bmatrix} a^2 & ab\cos\gamma & ac\cos\beta \\ ab\cos\gamma & b^2 & bc\cos\alpha \\ ac\cos\beta & bc\cos\alpha & c^2 \end{bmatrix}.$$

The matrix $\mathbf{G}$ is known as the *real space metric tensor* and is a symmetric, positive definite matrix. Matrix $\mathbf{A}$ will be recognized as its Cholesky decomposition. Its inverse, $\mathbf{G}^{-1} = (\mathbf{A}^T)^{-1}\mathbf{A}^{-1}$, is the *reciprocal space metric tensor*. The length of a vector, $\mathbf{h}$, expressed in reciprocal lattice units, may be expressed in reciprocal space units as $|\mathbf{d}^*| = (\mathbf{h}^T\mathbf{G}^{-1}\mathbf{h})^{1/2}$. Further consideration of the properties of this matrix must be deferred until we have discussed the properties of the reciprocal lattice.

Chapter 2

Symmetry of Finite Objects

Groups

A large proportion of the materials that exist in the solid state is crystalline, at least to some degree. Everything that is crystalline, along with many things that are not, possesses symmetry, which means that some operation —a rotation, a mirror reflection, an inversion, a translation, or some combination of these—brings it to a state that is indistinguishable from its original state. The mathematical treatment of symmetry makes extensive use of the theory of *groups*. A group is defined, quite arbitrarily, as a set of operations that combine[1] in such a way as to satisfy four postulates.

1. If two operations, a and b, are members of the group, their product, $c = ab$, is also a member of the group.
2. There exists one member of the group, designated E, such that $Ea = aE = a$, where a is any member of the group. E is known as the *identity operation*, or the *identity element*.
3. For every member, a, of the group there exists a member, a^{-1}, such that $aa^{-1} = a^{-1}a = E$. a^{-1} is known as the *inverse* of a.
4. Group operations in sequence obey the associative law: $a(bc) = (ab)c$.

Operations do not, in general, obey the commutative law, i.e., $ba \neq ab$, except that every member, as indicated in postulates 2 and 3, commutes with the identity element and with its own inverse.

Groups may be either finite or infinite. If a group is finite, it follows that each operation, if repeated a finite number of times, gives the identity element. A group that consists only of the identity element and powers of one other operation, a, so that its members are $E, a, a^2, \ldots, a^n (= E)$, is

[1]The process of combination is conventionally represented as multiplication, but it can actually be any of a large variety of operations.

known as the *cyclic group of order n* and is designated, in the notation introduced by Schönflies, C_n. The trivial group consisting only of E can be considered to be the cyclic group of order 1 and is, in fact, commonly designated C_1. A subset of the elements of a group that is itself a group is a *subgroup* of the group. In our example, if n is even, there is a group whose members are $E, a^2, a^4, \ldots, a^{n-2}$. This group is a *subgroup of index 2* of the original group. The original group is, correspondingly, a *supergroup* of the smaller group.

If a cyclic group, A, consisting of the members $E, a, a^2, \ldots, a^{n-1}$, and another cyclic group B, with members $E, b, b^2, \ldots, b^{m-1}$, such that $ab = ba$, are combined into a new group whose members are E, a, b, $ab(= ba)$, a^2b, etc., the resultant group, containing $m \times n$ members, is called the *direct product group* of A and B and is designated $A \times B$. It is a supergroup of both A and B.

A group is defined, ultimately, by a *multiplication table* which shows which members are products of which other members. As an example, we consider the so-called *four group* which is the direct product of two cyclic groups of order 2. Each entry in the table is the product of the member in the top row and that in the left-hand column.

	E	a	b	$c(= ab)$
E	E	a	b	c
a	a	E	c	b
b	b	c	E	a
c	c	b	a	E

It will be seen that each row and each column contain each member of the group.

Representations

The discussion of groups so far has been very general and very abstract. It is possible, however, to define sets of matrices which, according to the rules of matrix multiplication, form groups. For example, the set of matrices

$$\begin{pmatrix} 1 & 0 \\ 0 & 1 \end{pmatrix}, \begin{pmatrix} -1 & 0 \\ 0 & 1 \end{pmatrix}, \begin{pmatrix} 1 & 0 \\ 0 & -1 \end{pmatrix}, \begin{pmatrix} -1 & 0 \\ 0 & -1 \end{pmatrix},$$

obeys the relationships given in the multiplication table of the four group. The set of matrices is said to be a *representation* of the four group. In fact, because there is a one-to-one correspondence between the matrices and the group members, E, a, b, and c, the matrices form a *faithful representation* of the group. The set of 1×1 matrices (1), (-1), (1), (-1), also obeys the rules in the multiplication table, but there is no unique correspondence between the matrices and the members of the group. This set is a representation of the group, but not a faithful representation.

Note that the set of 2×2 matrices given above as a faithful representation can be constructed by placing the set of 1×1 matrices in the upper left corner and the alternative set (1), (1), (-1), (-1), which is also a representation, in the lower right corner, filling the rest of the elements in with zeros. A representation that can be built up by placing smaller ones along the diagonal, or which can be transformed to one by a similarity transformation, is said to be *reducible*. Correspondingly, one that cannot be so constructed is *irreducible*.

The number of rows (or columns, since they are square) in the matrices forming the representation is called the *dimension* of the representation. A remarkable theorem[2] states that the sum of the squares of the dimensions of all possible irreducible representations of a group is equal to the order of the group. In our example, the order of the group is four. We have given two one-dimensional representations. Every group has a one-dimensional representation formed by the matrix (1) repeated the number of times there are members of the group. There should, therefore, be one more one-dimensional representation. It is (1), (-1), (-1), (1).

Point Groups

Finite physical objects can exhibit various types of symmetry, all of which have the property that their operations leave at least one point unchanged. The sets of operations that transform the object to configurations indistinguishable from the original one form representations of a group and, because of the property of leaving a point unchanged, are called point groups. For three-dimensional objects the point groups are really three-dimensional, faithful representations of certain groups.

Before expanding further on the point groups we need a brief discussion of notation. Much of the theory of symmetry was worked out in the nineteenth century by Schönflies, who introduced a notation for labeling the groups, part of which, that for the cyclic groups, we have already seen. He considered these groups in terms of rotations around an axis through an angle equal to $2\pi/n$, so that n repetitions would produce an identity operation. Such operations are represented by the matrix

$$\mathbf{R} = \begin{pmatrix} \cos 2\pi/n & \sin 2\pi/n & 0 \\ -\sin 2\pi/n & \cos 2\pi/n & 0 \\ 0 & 0 & 1 \end{pmatrix},$$

assuming that the axis is parallel to x_3. This operation, as well as the point group it generates, is denoted C_n. Many subfields of materials science have

[2]A proof of this theorem appears in M. Hamermesh, *Group Theory and Its Application to Physical Problems*. Addison-Wesley Publishing Co., Reading, Massachusetts/Palo Alto/London, 1962, pp. 98–107. This book contains a very thorough exposition of group theory, but it is written from the point of view of particle physics and therefore contains much that is of little or no interest to the materials scientist.

had, at least until recently, to consider symmetry only in terms of point groups, with the result that, because of its historical precedence, the Schönflies notation has persisted and is deeply entrenched, particularly in such disciplines as atomic and molecular spectroscopy. However, with the advent of X-ray diffraction and, in the 1920s, the subsequent development of structural crystallography, it became necessary to make extensive use of the *space groups*, which describe the symmetry properties not of finite objects but of objects generated by the repeat of an infinite, space lattice. The space groups are derived from the point groups by adding the lattice translations and by combining with the point group operations translations that, when repeated a finite number of times, produce a lattice translation instead of the identity operation. Symbols for the space groups were produced by adding a superscript number to the Schönflies symbol for the corresponding point group, giving symbols of the form C_n^m.

Because there can be as many as 28 different space groups associated with a single point group, this notation conveys very little information to the user. The result was the development, by Hermann and Mauguin, of an alternative notation that is better adapted to producing a descriptive symbol for the space groups than is the Schönflies notation. For the pure (commonly called proper) rotation axes, the Hermann-Mauguin symbols substitute the number n, the order of the axis, for the symbol C_n. Thus C_2 becomes 2, C_3 becomes 3, etc. In the discussion that follows we shall give both the Schönflies and the Hermann-Mauguin symbols, which we shall refer to as H-M symbols for convenience, but we shall use the Hermann-Mauguin symbols exclusively for the space groups.

In addition to the proper rotations there is another class of symmetry operations known as *improper* rotations. These are represented by the product of the rotation axis matrix and either the inversion through the origin or a reflection across a plane perpendicular to the axis, represented, respectively, by

$$\begin{pmatrix} -1 & 0 & 0 \\ 0 & -1 & 0 \\ 0 & 0 & -1 \end{pmatrix} \quad \text{and} \quad \begin{pmatrix} 1 & 0 & 0 \\ 0 & 1 & 0 \\ 0 & 0 & -1 \end{pmatrix}.$$

The Schönflies system uses the second procedure and denotes the operator by S_n. The matrix representation is therefore

$$\mathbf{R} = \begin{pmatrix} \cos 2\pi/n & \sin 2\pi/n & 0 \\ -\sin 2\pi/n & \cos 2\pi/n & 0 \\ 0 & 0 & -1 \end{pmatrix}.$$

The H-M system uses the first procedure and denotes the operation by $\bar{n}$. The matrix representation is

$$\mathbf{R} = \begin{pmatrix} -\cos 2\pi/n & -\sin 2\pi/n & 0 \\ \sin 2\pi/n & -\cos 2\pi/n & 0 \\ 0 & 0 & -1 \end{pmatrix}.$$

It should be noted that S_4 is equivalent to $\bar{4}$, but S_3 is equivalent not to $\bar{3}$ but to $\bar{6}$, and S_6 is equivalent to $\bar{3}$. Both systems represent no symmetry at all, the group consisting of the identity operator only, by a rotation of 2π around an arbitrary axis. Thus the symbols are C_1 and 1. The inversion is represented in the H-M system by $\bar{1}$. It could be represented in the Schönflies system by S_2, but actually the special symbol C_i is used. Similarly, a mirror plane could be represented by S_1 or by $\bar{2}$, but it is actually represented in both systems by special symbols, C_s in Schönflies and m in H-M.

Because of the properties of space lattices, the only rotation axes, proper or improper, allowed in crystals are those of order 2, 3, 4, and 6. Molecules may have axes of order 5 and 7 or more, but the properties of the corresponding groups are sufficiently similar to those of the crystallographic groups that we shall not discuss them further here. Given the symmetry operations allowed in crystals there are 32 distinct, three-dimensional representations of groups. These are commonly referred to as the *32 crystallographic point groups*, but we should understand that they are actually group representations. Further, because some sets of these representations actually have identical multiplication tables—they are *isomorphic* —the number of distinct groups corresponding to these representations is only 18.

We shall give for each point group representation the following information: (1) the Schönflies symbol (or symbols, if more than one appears in common use); (2) the H-M symbol; (3) a stereographic representation[3] of points on a sphere that are equivalent under group operations; and (4) a group generator. This is a set of matrices that will, by forming all powers and products, produce all matrices in the representation.

In addition, for each set of isomorphic representations, we give a *character table*. The *character* of a group operation is just another name for the trace of the matrix representing that operation. Before describing the arrangement of the character tables, we must first discuss the concept of classes of operations. The traditional way to define a class is to state that if, for a pair of group operations, a and b, there exists another member of the group, c, such that $ac = cb$, then a and b belong to the same class. Perhaps a clearer picture of what this implies is obtained if we premultiply both sides of the equation by c^{-1}, giving $c^{-1}ac = b$. If we then represent these operations by matrices, **A**, **B**, and **C**, we can see that **A** and **B** are related by a similarity transformation. As we have seen, a similarity transformation leaves the trace invariant. Therefore the characters of all operations in a given class will be equal for any representation. A character table, therefore, can lump the operations in each class together, giving the character of the class as a whole.

Any representation can be expressed as a "sum" of one or more irreduci-

[3] An explanation of stereographic projections is given in Appendix B.

ble representations. Here the sum of two representations is the set of matrices formed by placing the corresponding matrices from each representation along the main diagonal and then filling the resulting matrix with zeros to make it square. The characters of any representation are therefore the sums of the characters of the irreducible representations that contribute to it. The character tables give the characters of all irreducible representations. We shall see (page 37) that these may, in many cases, be used to determine how a multidimensional representation may be reduced.

In presenting character tables there is a customary notation that is used to label the symmetry operators and another customary notation that is used to label the distinct, irreducible representations. Proper and improper rotations are designated by the Schönflies symbol for the cyclic groups they generate, C_n, or S_n. The inversion is designated by i and mirror planes by σ. Apostrophes may be used to distinguish operators of the same type that are in different classes, whereas subscripts or letters in parentheses may be used to designate the orientations of rotation axes (particularly twofold) and mirror planes.

One-dimensional representations are labeled by the letters A and B, with various sets of subscript numbers or apostrophes to distinguish similar ones. The letter E usually designates a two-dimensional representation. However, cyclic groups of order higher than 2 have one-dimensional representations generated by the complex nth roots of 1. The conjugate pairs of these are often linked and also labeled by E. Three-dimensional representations are designated by the letters F or T. Finally, for the representations of groups that contain the inversion center, the symbols for the irreducible representations will have the subscript g (for gerade) or u (for ungerade) depending on whether the character of the matrix corresponding to the inversion is positive or negative. Note that because the identity operation, E, is represented by the identity matrix, $\mathbf{I}$, the character of the identity operation is equal to the dimension of the irreducible representation. Note also that the number 1 is a 1×1 matrix that is a representation of *any* group. There is, therefore, for every group a one-dimensional, irreducible representation that associates the number 1 with every operation. This representation is called the *fully symmetric representation* and will always be labeled A, A', A_1, or A_{1g}. There is a system of sorts for the assignment of subscripts and apostrophes to the other representation labels, but it is not particularly informative, so we shall let the tables themselves serve as the definitions of these symbols.

There is a trivial group consisting of the identity only.
Symbols: $C_1 - 1$

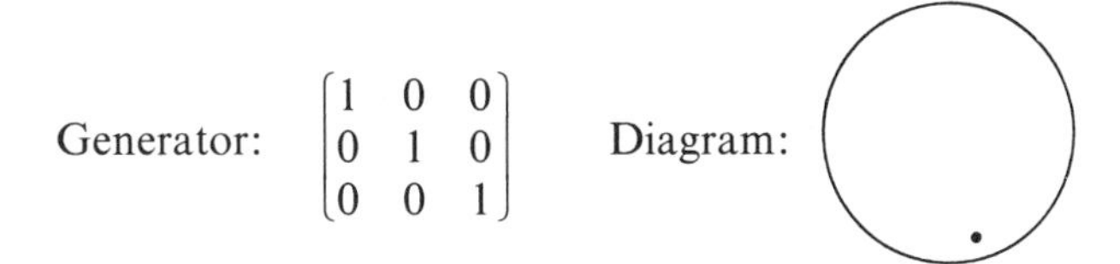

Character table:

1	E
A	1

There are three representations of the cyclic group of order 2.
Symbols: $C_i - \bar{1}$

Generator: $\begin{bmatrix} -1 & 0 & 0 \\ 0 & -1 & 0 \\ 0 & 0 & -1 \end{bmatrix}$ Diagram:

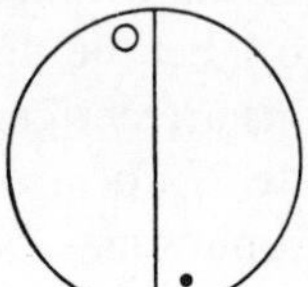

Symbols: $C_2 - 2$

Generator: $\begin{bmatrix} -1 & 0 & 0 \\ 0 & 1 & 0 \\ 0 & 0 & -1 \end{bmatrix}$ Diagram:

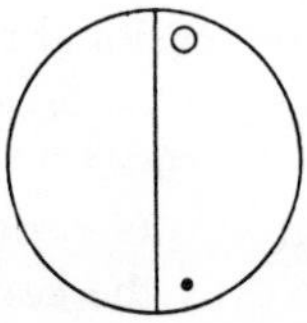

Symbols: $C_s - m$

Generator: $\begin{bmatrix} 1 & 0 & 0 \\ 0 & -1 & 0 \\ 0 & 0 & 1 \end{bmatrix}$ Diagram:

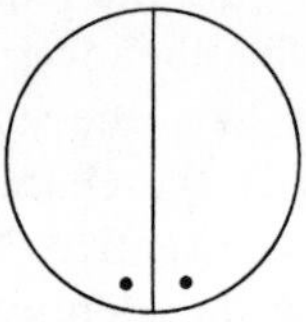

Character table:

$\bar{1}$	2	m	E	i, C_2, σ
A_g	A	A'	1	1
A_u	B	A''	1	-1

There are also three representations of the four group.
Symbols: $C_{2h} - 2/m$

Generator: $\begin{bmatrix} -1 & 0 & 0 \\ 0 & 1 & 0 \\ 0 & 0 & -1 \end{bmatrix}\begin{bmatrix} 1 & 0 & 0 \\ 0 & -1 & 0 \\ 0 & 0 & 1 \end{bmatrix}$ Diagram:

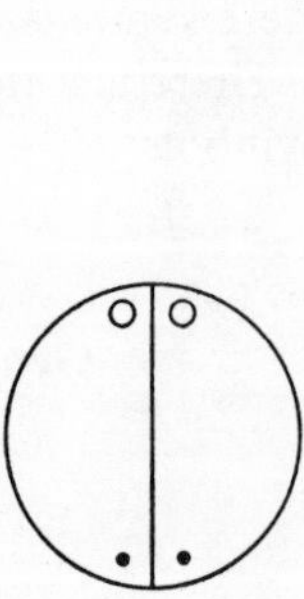

(Long-standing tradition has the unique direction in the point groups 2, m, and $2/m$ parallel to the y axis. *The International Tables for X-ray Crystallography* give both this setting and the alternative one with the unique direction parallel to the z axis, but the y setting is virtually universal in the literature.)

Symbols: $C_{2v} - mm2$

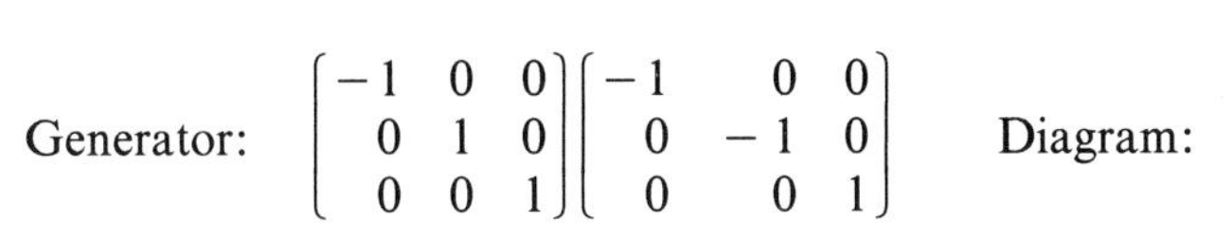

Generator: $\begin{pmatrix} -1 & 0 & 0 \\ 0 & 1 & 0 \\ 0 & 0 & 1 \end{pmatrix} \begin{pmatrix} -1 & 0 & 0 \\ 0 & -1 & 0 \\ 0 & 0 & 1 \end{pmatrix}$ Diagram:

Symbols: D_2 (or V) $- 222$

Generator: $\begin{pmatrix} -1 & 0 & 0 \\ 0 & -1 & 0 \\ 0 & 0 & 1 \end{pmatrix} \begin{pmatrix} -1 & 0 & 0 \\ 0 & 1 & 0 \\ 0 & 0 & -1 \end{pmatrix}$ Diagram:

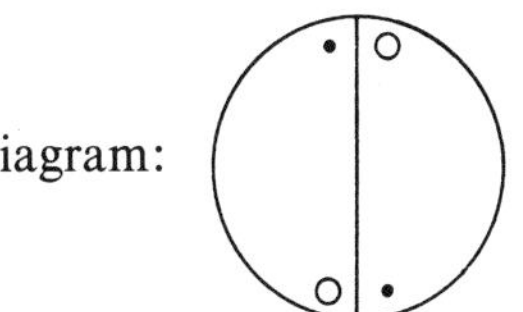

Character table:

$2/m$			E	C_2	i	σ_h
	$mm2$			C_2	$\sigma_v(xz)$	$\sigma_v(yz)$
		222		$C_2(z)$	$C_2(y)$	$C_2(x)$
A_g	A_1	A	1	1	1	1
A_u	A_2	B_1	1	1	-1	-1
B_g	B_1	B_2	1	-1	1	-1
B_u	B_2	B_3	1	-1	-1	1

There is one representation of the cyclic group of order 3.

Symbols: $C_3 - 3$

Generator: $\begin{pmatrix} -1/2 & -\sqrt{3}/2 & 0 \\ \sqrt{3}/2 & -1/2 & 0 \\ 0 & 0 & 1 \end{pmatrix}$ Diagram:

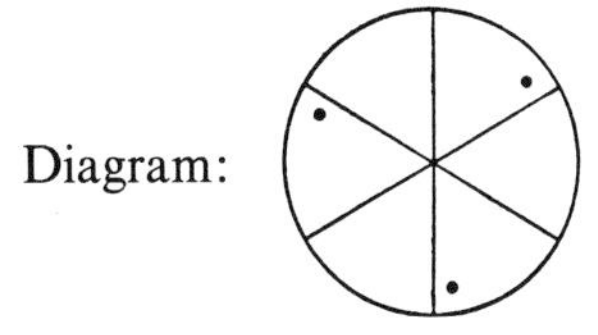

Character table:

3	E	C_3	C_3^2
A	1	1	1
E	1	$-1/2+(\sqrt{3}/2)i$	$-1/2-(\sqrt{3}/2)i$
	1	$-1/2-(\sqrt{3}/2)i$	$-1/2+(\sqrt{3}/2)i$

There are two representations of the cyclic group of order 4.
Symbols: $C_4 - 4$

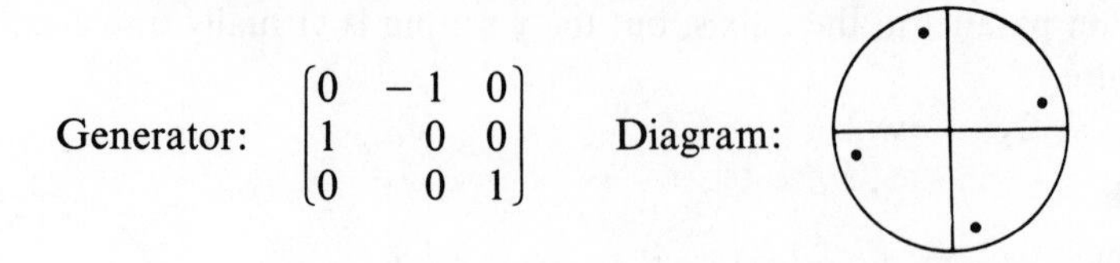

Generator: $\begin{bmatrix} 0 & -1 & 0 \\ 1 & 0 & 0 \\ 0 & 0 & 1 \end{bmatrix}$ Diagram:

Symbols: $S_4 - \bar{4}$

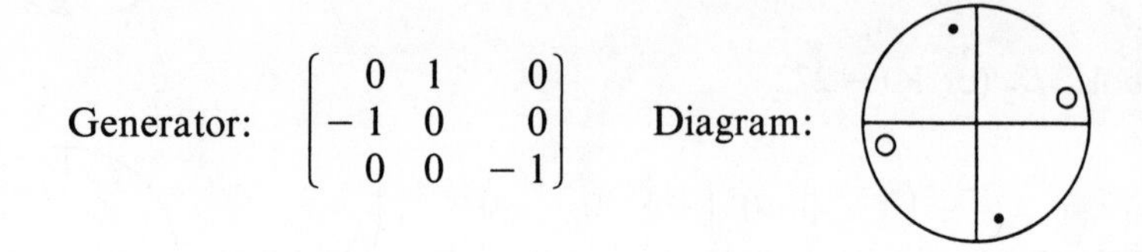

Generator: $\begin{bmatrix} 0 & 1 & 0 \\ -1 & 0 & 0 \\ 0 & 0 & -1 \end{bmatrix}$ Diagram:

Character table:

4	E	C_4	C_2	C_4^3
$\bar{4}$		S_4	C_2	S_4^3
A	1	1	1	1
B	1	-1	1	-1
E	1	i	-1	$-i$
	1	$-i$	-1	i

There are two distinct ways to combine the cyclic group of order 3 with the cyclic group of order 2. One of these is the direct product group, of which there are three representations.
Symbols S_6 (or C_{3i}) $- \bar{3}$

Generator: $\begin{bmatrix} 1/2 & \sqrt{3}/2 & 0 \\ -\sqrt{3}/2 & 1/2 & 0 \\ 0 & 0 & -1 \end{bmatrix}$ Diagram:

Symbols: $C_6 - 6$

Generator: $\begin{bmatrix} 1/2 & \sqrt{3}/2 & 0 \\ -\sqrt{3}/2 & 1/2 & 0 \\ 0 & 0 & 1 \end{bmatrix}$ Diagram:

Symbols: $C_{3h} - \bar{6}$

Generator: $\begin{pmatrix} -1/2 & -\sqrt{3}/2 & 0 \\ \sqrt{3}/2 & -1/2 & 0 \\ 0 & 0 & -1 \end{pmatrix}$ Diagram:

Character table: $[e = -1/2 + (\sqrt{3/2})i]$

$\bar{3}$			E	C_3	C_3^2	i	S_6^5	S_6
	6			C_3	C_3^2	C_2	C_6	C_6^5
		$\bar{6}$		C_3	C_3^2	σ_h	S_3	S_3^5
A_g	A	A'	1	1	1	1	1	1
E_g	E_2	E'	1	e	e^*	1	e	e^*
			1	e^*	e	1	e^*	e
A_u	B	A''	1	1	1	-1	-1	-1
E_u	E_1	E''	1	e	e^*	-1	$-e$	$-e^*$
			1	e^*	e	-1	$-e^*$	$-e$

In two other representations the elements of the cyclic group of order 2 do not commute with those of the cyclic group of order 3.
Symbols: $D_3 - 32$

Generator: $\begin{pmatrix} -1/2 & -\sqrt{3}/2 & 0 \\ \sqrt{3}/2 & -1/2 & 0 \\ 0 & 0 & 1 \end{pmatrix} \begin{pmatrix} 1 & 0 & 0 \\ 0 & -1 & 0 \\ 0 & 0 & -1 \end{pmatrix}$

Diagram:

Symbols: $C_{3v} - 3m$

Generator: $\begin{pmatrix} -1/2 & -\sqrt{3}/2 & 0 \\ \sqrt{3}/2 & -1/2 & 0 \\ 0 & 0 & 1 \end{pmatrix} \begin{pmatrix} 1 & 0 & 0 \\ 0 & -1 & 0 \\ 0 & 0 & 1 \end{pmatrix}$

Diagram:

Character table:

32 3m	E	$2C_3$ $2C_3$	$3C_2$ $3\sigma_v$
A_1	1	1	1
A_2	1	1	−1
E	2	−1	0

There is one representation that is a direct product of three cyclic groups or order 2.
Symbols: D_{2h} (or V_h) – *mmm*

Generators: $\begin{pmatrix} -1 & 0 & 0 \\ 0 & 1 & 0 \\ 0 & 0 & 1 \end{pmatrix} \begin{pmatrix} 1 & 0 & 0 \\ 0 & -1 & 0 \\ 0 & 0 & 1 \end{pmatrix} \begin{pmatrix} 1 & 0 & 0 \\ 0 & 1 & 0 \\ 0 & 0 & -1 \end{pmatrix}$

Diagram: 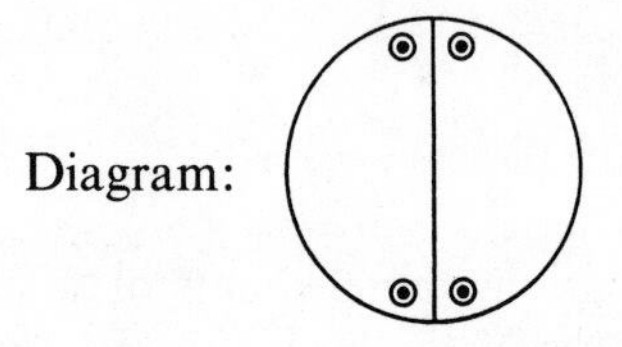

Character table:

mmm	E	$C_2(z)$	$C_2(y)$	$C_2(x)$	i	$\sigma(xy)$	$\sigma(xz)$	$\sigma(yz)$
A_g	1	1	1	1	1	1	1	1
B_{1g}	1	1	−1	−1	1	1	−1	−1
B_{2g}	1	−1	1	−1	1	−1	1	−1
B_{3g}	1	−1	−1	1	1	−1	−1	1
A_u	1	1	1	1	−1	−1	−1	−1
B_{1u}	1	1	−1	−1	−1	−1	1	1
B_{2u}	1	−1	1	−1	−1	1	−1	1
B_{3u}	1	−1	−1	1	−1	1	1	−1

There are two distinct ways to combine the cyclic group of order 4 with the cyclic group of order 2. One of these is the direct product group, which has one representation.
Symbols: D_{4h} – $4/m$

Generators: $\begin{pmatrix} 0 & 1 & 0 \\ -1 & 0 & 0 \\ 0 & 0 & 1 \end{pmatrix} \begin{pmatrix} 1 & 0 & 0 \\ 0 & 1 & 0 \\ 0 & 0 & -1 \end{pmatrix}$ Diagram:

Character table:

$4/m$	E	C_4	C_2	C_4^3	i	S_4^3	σ_h	S_4
A_g	1	1	1	1	1	1	1	1
B_g	1	-1	1	-1	1	-1	1	-1
E_g	1	i	-1	$-i$	1	i	-1	$-i$
	1	$-i$	-1	i	1	$-i$	-1	i
A_u	1	1	1	1	-1	-1	-1	-1
B_u	1	-1	1	-1	-1	1	-1	1
E_u	1	i	-1	$-i$	-1	$-i$	1	i
	1	$-i$	-1	i	-1	i	1	$-i$

In three other representations the elements of the cyclic group of order 4 and those of the group of order 2 do not commute.
Symbols: $D_4 - 422$

Generators: $\begin{pmatrix} 0 & 1 & 0 \\ -1 & 0 & 0 \\ 0 & 0 & 1 \end{pmatrix} \begin{pmatrix} 1 & 0 & 0 \\ 0 & -1 & 0 \\ 0 & 0 & -1 \end{pmatrix}$ Diagram:

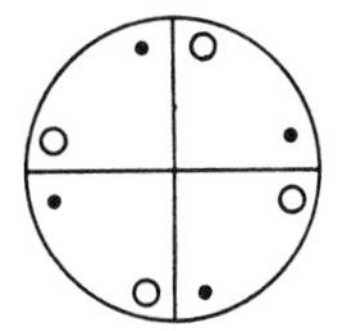

Symbols: $C_{4v} - 4mm$

Generators: $\begin{pmatrix} 0 & 1 & 0 \\ -1 & 0 & 0 \\ 0 & 0 & 1 \end{pmatrix} \begin{pmatrix} -1 & 0 & 0 \\ 0 & 1 & 0 \\ 0 & 0 & 1 \end{pmatrix}$ Diagram:

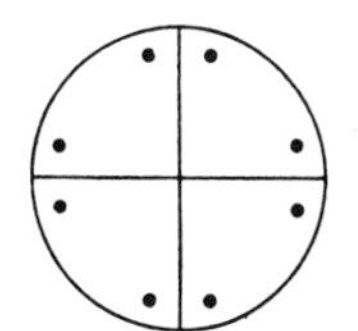

Symbols: $D_{2d} - \bar{4}2m$

Generators: $\begin{pmatrix} 0 & 1 & 0 \\ -1 & 0 & 0 \\ 0 & 0 & 1 \end{pmatrix} \begin{pmatrix} 1 & 0 & 0 \\ 0 & -1 & 0 \\ 0 & 0 & -1 \end{pmatrix}$ Diagram:

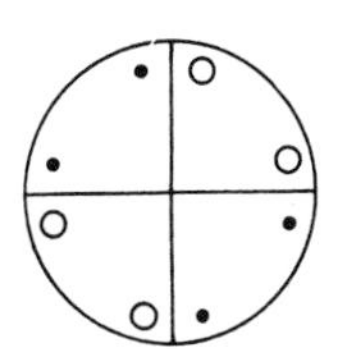

Character table:

422	E	$2C_4$	C_2	$2C_2'$	$2C_2''$
$4mm$		$2C_4$	C_2	$2\sigma_v$	$2\sigma_2$
$\bar{4}2m$		$2S_4$	C_2	$2C_2'$	$2\sigma_2$
A_1	1	1	1	1	1
A_2	1	1	1	-1	-1
B_1	1	-1	1	1	-1
B_2	1	-1	1	-1	1
E	2	0	-2	0	0

There are two distinct ways to combine the cyclic group of order 6 with the cyclic group of order 2. One of these is the direct product group, which has one representation.
Symbols: $C_{6h} - 6/m$

Generators: $\begin{pmatrix} 1/2 & \sqrt{3}/2 & 0 \\ -\sqrt{3}/2 & 1/2 & 0 \\ 0 & 0 & 1 \end{pmatrix} \begin{pmatrix} 1 & 0 & 0 \\ 0 & 1 & 0 \\ 0 & 0 & -1 \end{pmatrix}$

Diagram:

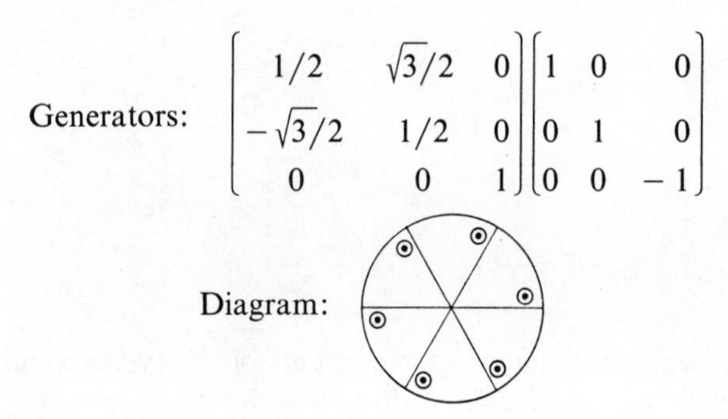

Character table: ($e = \exp 2\pi i/6$)

$6/m$	E	C_6	C_3	C_2	C_3^2	C_6^5	i	S_3^5	S_6^5	σ_h	S_6	S_3
A_g	1	1	1	1	1	1	1	1	1	1	1	1
B_g	1	-1	1	-1	1	-1	1	-1	1	-1	1	-1
E_{1g}	1	e	e^2	-1	e^{*2}	e^*	1	e	e^2	-1	e^{*2}	e^*
	1	e^*	e^{*2}	-1	e^2	e	1	e^*	e^{*2}	-1	e^2	e
E_{2g}	1	$-e^*$	e^{*2}	1	e^2	$-e$	1	$-e^*$	e^{*2}	1	e^2	$-e$
	1	$-e$	e^2	1	e^{*2}	$-e^*$	1	$-e$	e^2	1	e^{*2}	$-e^*$
A_u	1	1	1	1	1	1	-1	-1	-1	-1	-1	-1
B_u	1	-1	1	-1	1	-1	-1	1	-1	1	-1	1
E_{1u}	1	e	e^2	-1	e^{*2}	e^*	-1	$-e$	$-e^2$	1	$-e^{*2}$	$-e^*$
	1	e^*	e^{*2}	-1	e^2	e	-1	$-e^*$	$-e^{*2}$	1	$-e^2$	$-e$
E_{2u}	1	$-e^*$	e^{*2}	1	e^2	$-e$	-1	e^*	$-e^{*2}$	-1	$-e^2$	e
	1	$-e$	e^2	1	e^{*2}	$-e^*$	-1	e	$-e^2$	-1	$-e^{*2}$	e^*

In four other representations the elements of the cyclic group of order 6 and those of the group of order 2 do not commute.
Symbols: $D_{3d} - \bar{3}m$

Generators: $\begin{pmatrix} 1/2 & -\sqrt{3}/2 & 0 \\ \sqrt{3}/2 & 1/2 & 0 \\ 0 & 0 & -1 \end{pmatrix} \begin{pmatrix} -1 & 0 & 0 \\ 0 & 1 & 0 \\ 0 & 0 & 1 \end{pmatrix}$

Diagram:

Symbols: $C_{6v} - 6mm$

Generators: $$\begin{pmatrix} 1/2 & \sqrt{3}/2 & 0 \\ -\sqrt{3}/2 & 1/2 & 0 \\ 0 & 0 & 1 \end{pmatrix} \begin{pmatrix} -1 & 0 & 0 \\ 0 & 1 & 0 \\ 0 & 0 & 1 \end{pmatrix}$$

Diagram: 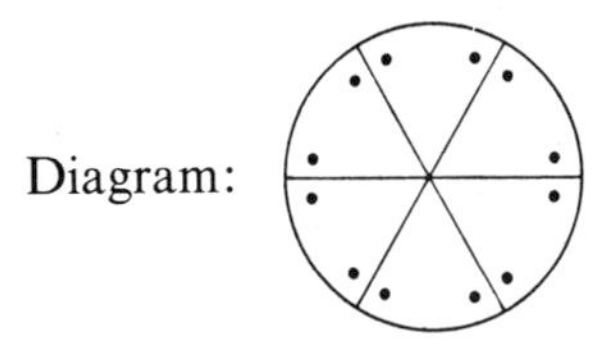

Symbols: $D_{3h} - \bar{6}m2$

Generators: $$\begin{pmatrix} -1/2 & -\sqrt{3}/2 & 0 \\ \sqrt{3}/2 & -1/2 & 0 \\ 0 & 0 & -1 \end{pmatrix} \begin{pmatrix} -1 & 0 & 0 \\ 0 & 1 & 0 \\ 0 & 0 & 1 \end{pmatrix}$$

Diagram: 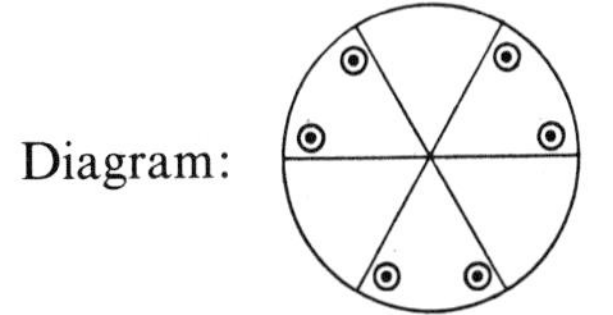

Symbols: $D_6 - 622$

Generators: $$\begin{pmatrix} 1/2 & \sqrt{3}/2 & 0 \\ -\sqrt{3}/2 & 1/2 & 0 \\ 0 & 0 & 1 \end{pmatrix} \begin{pmatrix} 1 & 0 & 0 \\ 0 & -1 & 0 \\ 0 & 0 & -1 \end{pmatrix}$$

Diagram: 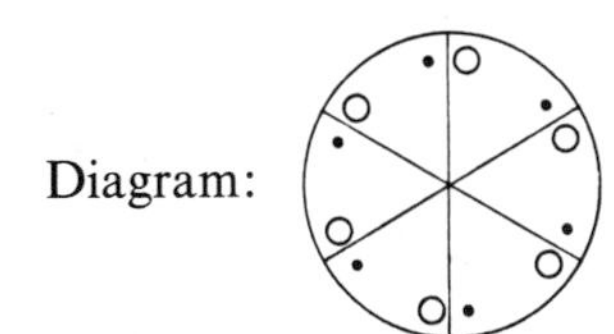

Character table:

$\bar{3}m$			E	$2S_6$	$2C_3$	i	$3C_2$	$3\sigma_2$
	$\bar{6}m2$			$2S_3$	$2C_3$	σ_h	$3C_2$	$3\sigma_v$
		$6mm$		$2C_6$	$2C_3$	C_2	$3\sigma_v$	$3\sigma_2$
		622		$2C_6$	$2C_3$	C_2	$3C_2'$	$3C_2''$
A_{1g}	A_1'	A_1	1	1	1	1	1	1
A_{2g}	A_2'	A_2	1	1	1	1	−1	−1
A_{1u}	A_1''	B_1	1	−1	1	−1	1	−1
A_{2u}	A_2''	B_2	1	−1	1	−1	−1	1
E_u	E''	E_1	2	1	−1	−2	0	0
E_g	E'	E_2	2	−1	−1	2	0	0

Two more representations are representations of direct products of the cyclic group of order 2 with previously derived representations.
Symbols: $D_{4h} - 4/mmm$

$$\text{Generators:}\quad \begin{pmatrix} 0 & 1 & 0 \\ -1 & 0 & 0 \\ 0 & 0 & 1 \end{pmatrix} \begin{pmatrix} 1 & 0 & 0 \\ 0 & 1 & 0 \\ 0 & 0 & -1 \end{pmatrix} \begin{pmatrix} -1 & 0 & 0 \\ 0 & 1 & 0 \\ 0 & 0 & 1 \end{pmatrix}$$

Diagram:

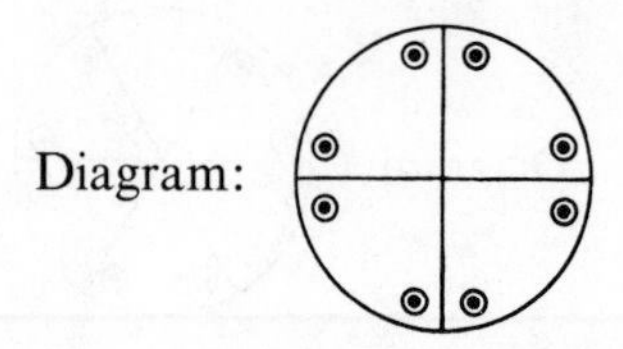

Character table:

$4/mmm$	E	$2C_4$	C_2	$2C_2'$	$2C_2''$	i	$2S_4$	σ_h	$2\sigma_v$	$2\sigma_d$
A_{1g}	1	1	1	1	1	1	1	1	1	1
A_{2g}	1	1	1	−1	−1	1	1	1	−1	−1
B_{1g}	1	−1	1	1	−1	1	−1	1	1	−1
B_{2g}	1	−1	1	−1	1	1	−1	1	−1	1
E_g	2	0	−2	0	0	2	0	−2	0	0
A_{1u}	1	1	1	1	1	−1	−1	−1	−1	−1
A_{2u}	1	1	1	−1	−1	−1	−1	−1	1	1
B_{1u}	1	−1	1	1	−1	−1	1	−1	−1	1
B_{2u}	1	−1	1	−1	1	−1	1	−1	1	−1
E_u	2	0	−2	0	0	−2	0	2	0	0

Symbols: $D_{6h} - 6/mmm$

Generators: $\begin{pmatrix} 1/2 & \sqrt{3}/2 & 0 \\ -\sqrt{3}/2 & 1/2 & 0 \\ 0 & 0 & 1 \end{pmatrix} \begin{pmatrix} 1 & 0 & 0 \\ 0 & 1 & 0 \\ 0 & 0 & -1 \end{pmatrix} \begin{pmatrix} -1 & 0 & 0 \\ 0 & 1 & 0 \\ 0 & 0 & 1 \end{pmatrix}$

Diagram:

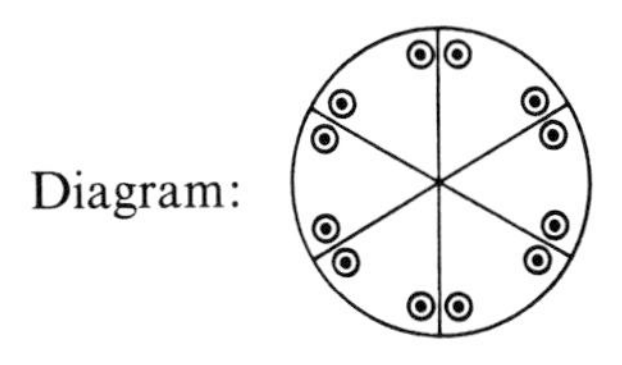

Character table:

$6/mmm$	E	$2C_6$	$2C_3$	C_2	$3C_2'$	$3C_2''$	i	$2S_3$	$2S_6$	σ_h	$3\sigma_d$	$3\sigma_v$
A_{1g}	1	1	1	1	1	1	1	1	1	1	1	1
A_{2g}	1	1	1	1	−1	−1	1	1	1	1	−1	−1
B_{1g}	1	−1	1	−1	1	−1	1	−1	1	−1	1	−1
B_{2g}	1	−1	1	−1	−1	1	1	−1	1	−1	−1	1
E_{1g}	2	1	−1	−2	0	0	2	1	−1	−2	0	0
E_{2g}	2	−1	−1	2	0	0	2	−1	−1	2	0	0
A_{1u}	1	1	1	1	1	1	−1	−1	−1	−1	−1	−1
A_{2u}	1	1	1	1	−1	−1	−1	−1	−1	−1	1	1
B_{1u}	1	−1	1	−1	1	−1	−1	1	−1	1	−1	1
B_{2u}	1	−1	1	−1	−1	1	−1	1	−1	1	1	−1
E_{1u}	2	1	−1	−2	0	0	−2	−1	1	2	0	0
E_{2u}	2	−1	−1	2	0	0	−2	1	1	−2	0	0

The remaining five point group representations are known as the *cubic point groups*, and all are characterized by a combination of one of the previously described representations with a threefold rotation about an axis forming an angle of $\cos^{-1}(1/\sqrt{3})[= 54° 44']$ with each of the coordinate axes.

Symbols: $T - 23$

Generators: $\begin{pmatrix} -1 & 0 & 0 \\ 0 & -1 & 0 \\ 0 & 0 & 1 \end{pmatrix} \begin{pmatrix} 0 & 1 & 0 \\ 0 & 0 & 1 \\ 1 & 0 & 0 \end{pmatrix}$ Diagram:

Character table: ($e = \exp 2\pi i/3$)

23	E	$4C_3$	$4C_3^2$	$3C_2$
A	1	1	1	1
E	1	e	e^*	1
	1	e^*	e	1
T	3	0	0	-1

Symbols: $T_h - m3$

Generators: $\begin{bmatrix} 1 & 0 & 0 \\ 0 & 1 & 0 \\ 0 & 0 & -1 \end{bmatrix} \begin{bmatrix} 0 & 1 & 0 \\ 0 & 0 & 1 \\ 1 & 0 & 0 \end{bmatrix}$ Diagram:

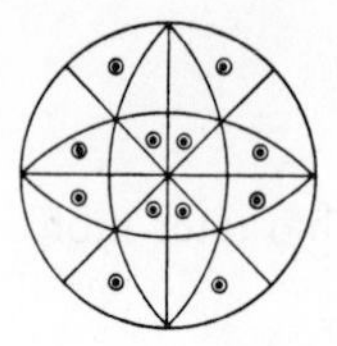

Character table: ($e = \exp 2\pi i/3$)

$m3$	E	$4C_3$	$4C_3^2$	$3C_2$	i	$4S_6^5$	$4S_6$	$3\sigma_h$
A_g	1	1	1	1	1	1	1	1
E_g	1	e	e^*	1	1	e	e^*	1
	1	e^*	e	1	1	e^*	e	1
T_g	3	0	0	-1	3	0	0	-1
A_u	1	1	1	1	-1	-1	-1	-1
E_u	1	e	e^*	1	-1	$-e$	$-e^*$	-1
	1	e^*	e	1	-1	$-e^*$	$-e$	-1
T_u	3	0	0	-1	-3	0	0	1

Symbols: $T_d - \bar{4}3m$

Generators: $\begin{bmatrix} 0 & 1 & 0 \\ -1 & 0 & 0 \\ 0 & 0 & -1 \end{bmatrix} \begin{bmatrix} 0 & 1 & 0 \\ 0 & 0 & 1 \\ 1 & 0 & 0 \end{bmatrix}$ Diagram:

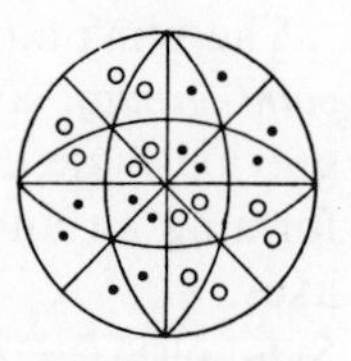

Symbols: $O - 432$

Generators: $\begin{bmatrix} 0 & 1 & 0 \\ -1 & 0 & 0 \\ 0 & 0 & 1 \end{bmatrix} \begin{bmatrix} 0 & 1 & 0 \\ 0 & 0 & 1 \\ 1 & 0 & 0 \end{bmatrix}$ Diagram:

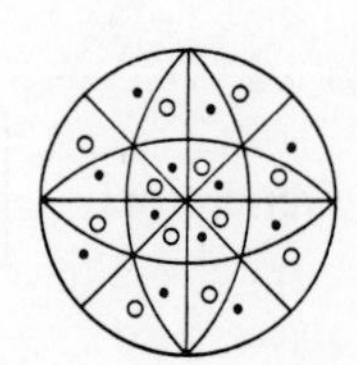

Character table:

$\bar{4}3m$ 432	E	$8C_3$ $8C_3$	$3C_2$ $3C_2$	$6S_4$ $6C_4$	$6\sigma_d$ $6C_2'$
A_1	1	1	1	1	1
A_2	1	1	1	-1	-1
E	2	-1	2	0	0
T_1	3	0	-1	1	-1
T_2	3	0	-1	-1	1

Symbols: $O_h - m3m$

Generators: $\begin{bmatrix} 0 & 1 & 0 \\ -1 & 0 & 0 \\ 0 & 0 & 1 \end{bmatrix} \begin{bmatrix} 0 & 1 & 0 \\ 0 & 0 & 1 \\ 1 & 0 & 0 \end{bmatrix} \begin{bmatrix} 0 & 1 & 0 \\ 1 & 0 & 0 \\ 0 & 0 & 1 \end{bmatrix}$

Diagram: 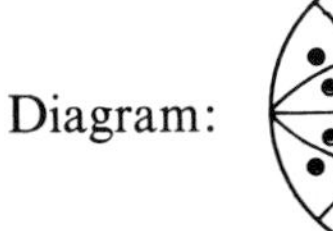

Character table:

$m3m$	E	$8C_3$	$3C_2$	$6C_4$	$6C_2'$	i	$8S_6$	$3\sigma_h$	$6S_4$	$6\sigma_d$
A_{1g}	1	1	1	1	1	1	1	1	1	1
A_{2g}	1	1	1	-1	-1	1	1	1	-1	-1
E_g	2	-1	2	0	0	2	-1	2	0	0
T_{1g}	3	0	-1	1	-1	3	0	-1	1	-1
T_{2g}	3	0	-1	-1	1	3	0	-1	-1	1
A_{1u}	1	1	1	1	1	-1	-1	-1	-1	-1
A_{2u}	1	1	1	-1	-1	-1	-1	-1	1	1
E_u	2	-1	2	0	0	-2	1	-2	0	0
T_{1u}	3	0	-1	1	-1	-3	0	1	-1	1
T_{2u}	3	0	-1	-1	1	-3	0	1	1	-1

Note that, in most of these cases, the choice of group generators is not unique. However, the H-M symbol always designates a sufficient (although, in a few cases, redundant) set. In each case we have given here the generator set that corresponds to the H-M symbol for the representation.

The character tables may be used to determine which irreducible representations contribute to a reducible one. To see how this is done let us consider a simple example, the point group representation 222. The four matrices of the representation are

$$\begin{bmatrix} 1 & 0 & 0 \\ 0 & 1 & 0 \\ 0 & 0 & 1 \end{bmatrix}, \begin{bmatrix} 1 & 0 & 0 \\ 0 & -1 & 0 \\ 0 & 0 & -1 \end{bmatrix}, \begin{bmatrix} -1 & 0 & 0 \\ 0 & 1 & 0 \\ 0 & 0 & -1 \end{bmatrix} \text{ and } \begin{bmatrix} -1 & 0 & 0 \\ 0 & -1 & 0 \\ 0 & 0 & 1 \end{bmatrix}.$$

The characters are 3, -1, -1, and -1. Let us designate by n_1, n_2, n_3, and n_4 the number of times the irreducible representations A, B_1, B_2, and B_3 each contribute to the representation. Since the trace of each matrix is the sum of the traces of the submatrices along the main diagonal, we can write the four equations

$$n_1 + n_2 + n_3 + n_4 = 3,$$

$$n_1 - n_2 - n_3 + n_4 = -1,$$

$$n_1 - n_2 + n_3 - n_4 = -1,$$

$$n_1 + n_2 - n_3 - n_4 = -1,$$

for which the solution is $n_1 = 0$, $n_2 = n_3 = n_4 = 1$. Thus the representation contains B_1, B_2, and B_3 once each, and does not contain A at all. Similarly, we can determine that the representation $2/m$ contains A_u once and B_u twice, whereas $mm2$ contains A_1, B_1, and B_2 once each.

Subgroups and Supergroups of the Point Groups. Figure 2.1 is a chart showing subgroup and supergroup relationships among the point groups. The highest order groups are on top, and each group is connected to its supergroups above and its subgroups below by lines. These relationships are very useful in studies of related structures, as in phase transitions and order-disorder.

Basis Functions

If we consider a representation of dimension n and consider that a set of functions $u_1, u_2, \ldots, u_n$, are the elements of an n-dimensional vector, then the vector that results from multiplying this vector by one of the matrices constituting the representation will be composed of linear combinations of the functions, u_i. These functions are said to be a set of *basis functions* for the representation. If the representation is reducible, it is possible to find linear combinations of the functions u_i that form smaller sets of functions that transform among themselves as a result of the operations of the group.

We have see that the 32 crystallographic "point groups" are actually three-dimensional representations (most of them reducible) of 18 distinct groups. In considering the symmetry of finite objects in real space, the basis functions for these representations are the components of the displacement

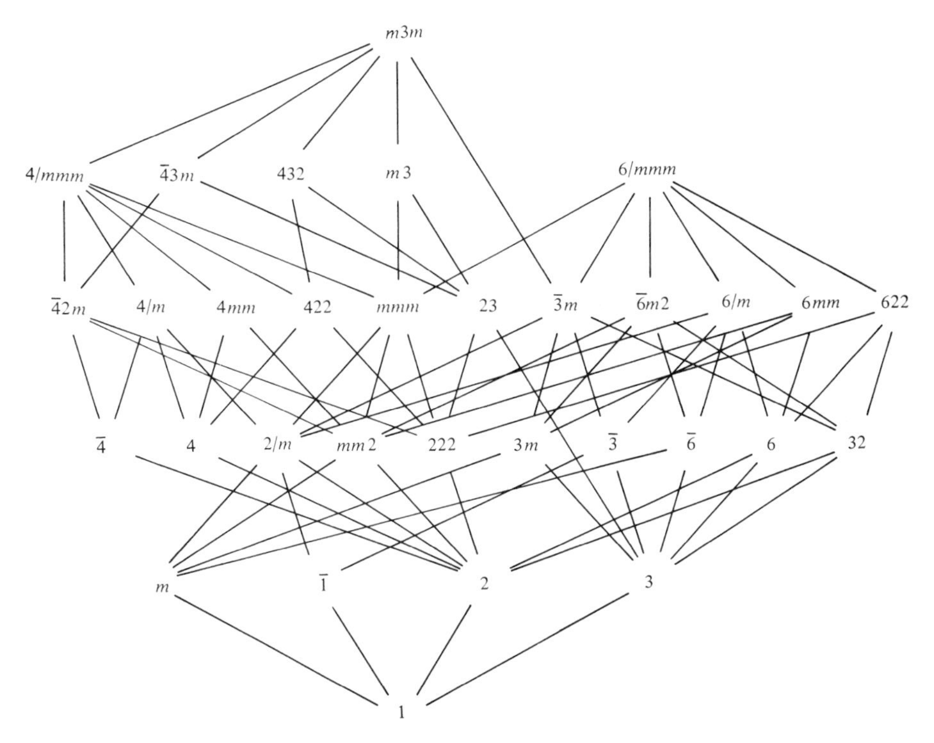

Fig. 2.1. Chart showing the subgroup and supergroup relationships for the crystallographic point groups.

of a point from the origin of an orthonormal, cartesian coordinate system: $u_1 = x$, $u_2 = y$, and $u_3 = z$. In the noncubic point groups the representation is reducible, and these functions, individually or in pairs, are the basis functions for various irreducible representations. In those point groups that have pairs of irreducible representations in which some of the characters are complex and conjugate to one another, a pair of coordinate displacements is usually a basis set for a real, two-dimensional representation that is reducible only by introducing complex elements. For example, the cyclic group of order 4 can be represented by the sum of the two representations designated E:

$$\begin{pmatrix} 1 & 0 \\ 0 & 1 \end{pmatrix}, \begin{pmatrix} i & 0 \\ 0 & -i \end{pmatrix}, \begin{pmatrix} -1 & 0 \\ 0 & -1 \end{pmatrix}, \text{ and } \begin{pmatrix} -i & 0 \\ 0 & i \end{pmatrix}.$$

If we perform similarity transformations using the pair of unitary matrices

$$\mathbf{T} = \begin{bmatrix} 1/\sqrt{2} & i/\sqrt{2} \\ i/\sqrt{2} & 1/\sqrt{2} \end{bmatrix} \text{ and } \mathbf{T}^{-1} = \begin{bmatrix} 1/\sqrt{2} & -i/\sqrt{2} \\ -i/\sqrt{2} & 1/\sqrt{2} \end{bmatrix},$$

on all of these matrices we obtain the real representation

$$\begin{pmatrix} 1 & 0 \\ 0 & 1 \end{pmatrix}, \begin{pmatrix} 0 & 1 \\ -1 & 0 \end{pmatrix}, \begin{pmatrix} -1 & 0 \\ 0 & -1 \end{pmatrix}, \text{ and } \begin{pmatrix} 0 & -1 \\ 1 & 0 \end{pmatrix}.$$

x and y are basis functions for this representation. z is a basis function for irreducible representation A in group 4 and for irreducible representation B in group $\bar{4}$.

Functions of x, y, and z may be basis functions for representations other than the one corresponding to the displacements. The squared distance from the origin, $x^2 + y^2 + z^2$, is always a basis functions for the fully symmetric representation. In groups possessing a center of symmetry x, y, and z are basis functions for u (ungerade or odd) representations, and x^2, y^2, and z^2 are basis functions for the corresponding g (gerade or even) representations. As a further example, in the cubic group 432 x, y, and z are basis functions for representation T. The pair of functions $u_1 = (\sqrt{3}/2) \cdot (x^2 - y^2)$ and $u_2 = z^2 - (1/2)(x^2 + y^2)$ are basis functions for the two-dimensional representation, E.

Chapter 3

Symmetry of Infinitely Repeated Patterns

Bravais Lattices

A *space lattice* is an infinite set of points[1] defined by the relationship $\mathbf{d} = n_1\mathbf{a} + n_2\mathbf{b} + n_3\mathbf{c}$. $\mathbf{a}$, $\mathbf{b}$, and $\mathbf{c}$ are vectors such that the matrix

$$\mathbf{V} = \begin{pmatrix} a_1 & a_2 & a_3 \\ b_1 & b_2 & b_3 \\ c_1 & c_2 & c_3 \end{pmatrix}$$

is nonsingular, and n_1, n_2, and n_3 are integers. $\mathbf{d}$ is a vector relating pairs of points.

If the origin and the point $(n_1\mathbf{a}, n_2\mathbf{b}, n_3\mathbf{c})$ are points of the lattice, then the point $(-n_1\mathbf{a}, -n_2\mathbf{b}, -n_3\mathbf{c})$ is one also. Thus a lattice is always centrosymmetric, but may also have more symmetry. If $\mathbf{d}$ is a vector between points of the lattice, and $\mathbf{A}$ is an operation of one of the point groups, then the vector $\mathbf{d}' = \mathbf{A}\mathbf{d}$ may also be a vector of the lattice. There are 11 point group representations that contain centers of symmetry, but because rotation axes combined with centers of symmetry often generate mirror planes, all lattices have the symmetry of one of the seven point groups $\bar{1}$, $2/m$, mmm, $\bar{3}m$, $4/mmm$, $6/mmm$, and $m3m$. These point groups define the seven *crystal systems*, which are designated, in the same order, *triclinic, monoclinic, orthorhombic, trigonal,*[2] *tetragonal, hexagonal,* and *cubic*.

[1]It should be emphasized at this time that a lattice is a *mathematical* concept, an infinite array of points fixed in space. A lattice does not vibrate, and it has *no* physical properties whatsoever! Thus, such deeply entrenched expressions as "lattice vibrations" and "lattice dynamics" are misuses of the language. The usage will undoubtedly continue unchallenged by editors and referees into the indefinite future, but it would be a good thing if the materials science community would find less self-contradictory expressions.

[2]Some trigonal space groups are based on a hexagonal lattice.

Any three noncoplanar lattice translations may be used to define a *unit cell*. If the set of translations is the shortest possible such set, or if it defines a cell with a volume equal to the volume of the shortest set, the unit cell is said to be primitive and is designated P. In a number of cases it is more convenient to define the unit cell with respect to lattice translations that are related in some way to symmetry operations—parallel to rotation axes or lying in mirror planes. This may require a unit cell whose volume is two, three, or four times the volume of the primitive cell. If the midpoint of the body diagonal of the unit cell is also a point of the lattice, the lattice is said to be *body-centered* and is designated I. If the midpoint of the diagonal of a single face of the unit cell is a lattice point, the cell is *single-face-centered*. Single-face centering is conventionally designated C, meaning that the centered face *does not* contain the **c** crystallographic axis, but, particularly if the system is orthorhombic, the designations A and B, indicating that the centered face does not contain the **a** or **b** axis, may appear in the literature. If the midpoints of all faces of the unit cell are points of the lattice, the lattice is said to be *all-face-centered*, or simply *face-centered*, and is designated F. Finally, it is usually convenient to describe the lattice whose point group symmetry is $\bar{3}m$, the rhombohedral or R lattice, by a hexagonal unit cell whose volume is three times that of the primitive cell. Taking account of symmetry and the various centering schemes, there are 14 distinct space lattices. This fact was established in the middle of the nineteenth century by A. Bravais, and the 14 lattices are therefore known as *Bravais lattices*. They are illustrated in Figure 3.1.

Even trained crystallographers often become confused when confronted with the R lattice described on hexagonal axes. This is because there are two distinct ways to fit a primitive rhombohedral unit cell into a hexagonal unit cell. These are known as the *obverse* and *reverse* settings of the rhombohedron. Although they are identical lattices, both are shown in the figure. The obverse setting has lattice points at fractional coordinates $\frac{2}{3}, \frac{1}{3}, \frac{1}{3}$ and $\frac{1}{3}, \frac{2}{3}, \frac{2}{3}$ in the hexagonal unit cell, whereas the reverse setting has lattice points at $\frac{1}{3}, \frac{2}{3}, \frac{1}{3}$ and $\frac{2}{3}, \frac{1}{3}, \frac{2}{3}$. The conditions for Bragg reflections to be allowed are $-h + k + l = 3n$ for the obverse setting and $h - k + l = 3n$ for the reverse setting.

Space Groups

The displacements defining a space lattice obey all of the conditions required for a group if the operation is vector addition. The null vector is the identity operation. The inverse of vector addition is vector subtraction, all possible vector sums lead to points of the lattice, and the order in which the operations are performed is immaterial. A space lattice is therefore a representation of an infinite group known as the *translation group*. In our discussion of point symmetry we saw that the operations, proper and improper rotations and reflections, were required to be cyclic—$\mathbf{A}^n = \mathbf{I}$, where n is 1, 2, 3, 4, or 6. If we consider operations in space we can define

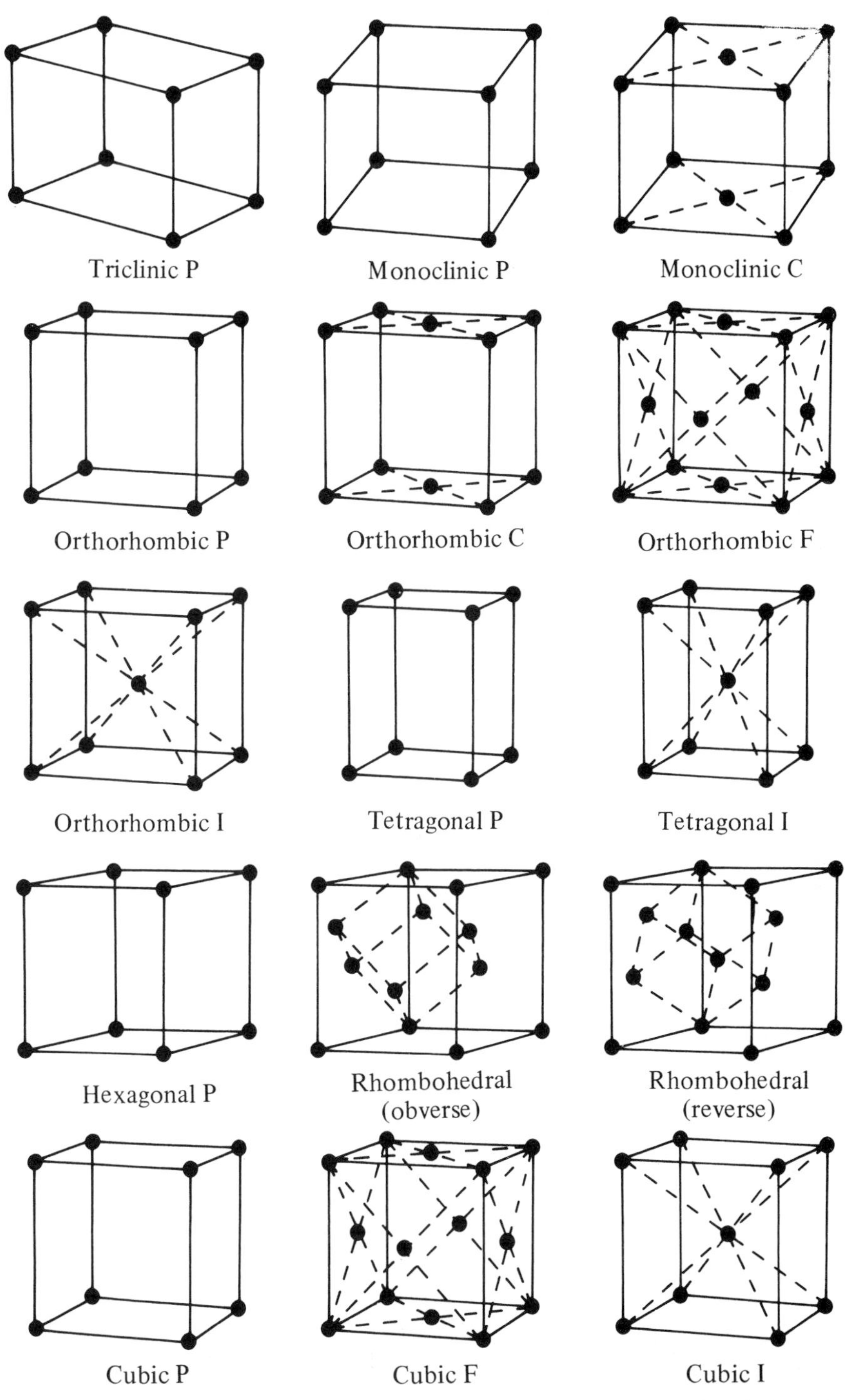

Fig. 3.1. The 14 Bravais lattices. The rhombohedral obverse and the rhombohedral reverse are identical lattices. Both are shown to illustrate their relationships to a hexagonal superlattice.

linear operators of the form $\mathbf{Ox} = \mathbf{Ax} + \mathbf{b}$, where $\mathbf{A}$ is a matrix and $\mathbf{b}$ is a vector of the same dimension as $\mathbf{x}$. If $\mathbf{O}^n\mathbf{x} = \mathbf{x} + \mathbf{a}$, where $\mathbf{a}$ is a translation vector of the lattice, then the operators $\mathbf{O}$ define a group that is a supergroup of the translation group.

If the vector $\mathbf{b}$ is the null vector, then the operations are those that define the point groups. The possibility of including lattice translations allows us to define two new types of symmetry operation known as *screw axes* and *glide planes*. In a screw axis a rotation, $\mathbf{A}$, is combined with a translation, $\mathbf{b}$, parallel to the rotation axis such that, if $\mathbf{A}^n = \mathbf{I}$, then $n\mathbf{b}$ is a lattice translation.

For example, a twofold screw axis parallel to the y axis corresponds to the linear operation

$$\begin{pmatrix} x' \\ y' \\ z' \end{pmatrix} = \begin{pmatrix} -1 & 0 & 0 \\ 0 & 1 & 0 \\ 0 & 0 & -1 \end{pmatrix} \begin{pmatrix} x \\ y \\ z \end{pmatrix} + \begin{pmatrix} 0 \\ 1/2 \\ 0 \end{pmatrix} = \begin{pmatrix} -x \\ y + 1/2 \\ -z \end{pmatrix}.$$

Repeating this operation, we obtain

$$\begin{pmatrix} x'' \\ y'' \\ z'' \end{pmatrix} = \begin{pmatrix} -1 & 0 & 0 \\ 0 & 1 & 0 \\ 0 & 0 & -1 \end{pmatrix} \begin{pmatrix} x' \\ y' \\ z' \end{pmatrix} + \begin{pmatrix} 0 \\ 1/2 \\ 0 \end{pmatrix} = \begin{pmatrix} x \\ y \\ z \end{pmatrix} + \begin{pmatrix} 0 \\ 1 \\ 0 \end{pmatrix},$$

which corresponds to a lattice translation in the y direction.

A glide plane arises when a mirror operation is combined with a vector, $\mathbf{b}$, parallel to the mirror plane, and $2\mathbf{b}$ is a lattice translation.

A glide plane perpendicular to the y axis with a translation in the z direction corresponds to the linear operation

$$\begin{pmatrix} x' \\ y' \\ z' \end{pmatrix} = \begin{pmatrix} 1 & 0 & 0 \\ 0 & -1 & 0 \\ 0 & 0 & 1 \end{pmatrix} \begin{pmatrix} x \\ y \\ z \end{pmatrix} + \begin{pmatrix} 0 \\ 0 \\ 1/2 \end{pmatrix} = \begin{pmatrix} x \\ -y \\ z + 1/2 \end{pmatrix}.$$

Repeating this operation, we obtain

$$\begin{pmatrix} x'' \\ y'' \\ z'' \end{pmatrix} = \begin{pmatrix} 1 & 0 & 0 \\ 0 & -1 & 0 \\ 0 & 0 & 1 \end{pmatrix} \begin{pmatrix} x' \\ y' \\ z' \end{pmatrix} + \begin{pmatrix} 0 \\ 0 \\ 1/2 \end{pmatrix} = \begin{pmatrix} x \\ y \\ z \end{pmatrix} + \begin{pmatrix} 0 \\ 0 \\ 1 \end{pmatrix},$$

which corresponds to a lattice translation in the z direction.

Making use of this set of symmetry operations, combined with the 14 Bravais lattices, Schönflies, Fedorov, and Barlow, independently, and more or less at the same time, showed that there are 230 distinct arrangements. These arrangements are known as *space groups*.

The space groups and their properties are listed in great detail in the *International Tables for X-ray Crystallography*, and it would be pointless to attempt to reproduce that extensive information. The tables are, however, not at all self-explanatory to someone who has not had extensive training in crystallography, and they can be confusing even to those who have. We shall therefore give a summary of this information and an explanation of how it may be interpreted.

The notations for labeling the space groups, both those of Schönflies and of Hermann and Mauguin, are based on extensions of point-group notation to include screw axes, glide planes, and lattice translations. The Schönflies notation is formed by attaching to the point-group symbol a superscript indicating the order in which he derived the space groups associated with each point group. Since no one can reproduce the historical order of Schönflies' reasoning process, it is impossible to determine which space group attaches to which symbol without reference to the *International Tables* or an equivalent list. The H-M symbols were developed in an attempt to produce a notation that would allow each space group to be labeled by a unique symbol that would enable the group to be derived without reference to the tables. Except for a small number of pairs of groups that differ from one another only in whether or not axes intersect in space, this attempt was largely successful.

An H-M space-group symbol begins with a capital letter designating the lattice type, P, C, I, F, or R. A or B may appear, in the literature but not in the tables, when there is some reason to use a nonstandard designation (setting) for a space group based on a single-face-centered lattice. The lattice symbol is followed by a modified point-group symbol. A rotation axis is designated, as before, by a number, with a bar over it if the rotation is improper. A mirror plane is again designated by the letter *m*. Screw axes are designated by the number, followed by a subscript number giving the numerator of the fraction of a lattice translation that accompanies a *right-handed* rotation. (If a right-handed screw is driven by a screwdriver held in the palm of the right hand and rotated in the direction of the fingers, the screw will advance in the direction of the thumb.) The denominator is the order of the rotation. Thus 6_2 designates a rotation of 60° accompanied by a translation of $\frac{1}{3}$ of a lattice translation in the right-handed-screw direction. 6_4 would designate a translation of $\frac{2}{3}$ of a lattice translation in the right-handed direction. Note, however, that a $\frac{2}{3}$ translation in the positive direction is equivalent to a $\frac{1}{3}$ translation in the negative direction. 6_4, therefore, may be considered to designate the left-handed screw that is a mirror image of 6_2.

Glide planes are represented by a letter giving the direction in the lattice in which a translation equal to $\frac{1}{2}$ of the lattice translation is to accompany the mirror reflection. Thus *a*, *b*, and *c* designate translations of $\frac{1}{2}$ the **a**, **b**, and **c** translations of the lattice, respectively. *n* designates a translation of $\frac{1}{2}$ of a face diagonal, and *d* designates a translation of $\frac{1}{2}$ the distance from the origin of the unit cell to a face or body center.

The information in the *International Tables for X-ray Crystallography* is given in a very condensed form.[3] On the theory that the best way of explaining something is to analyze an example, we shall discuss Figure 3.2, which is a facsimile of a fairly typical page from the book.

In the upper-right-hand corner the H-M symbol for the space group appears in large type, with the Schönflies symbol immediately below it. (This is a right-hand page. If it were a left-hand page the corresponding symbols would appear in the upper-left-hand corner. This makes it easy to find a particular space group by riffling through the pages with the thumb of one hand.) Across the top of the page, in smaller type, are the crystal system, the H-M, point-group symbol for the corresponding point group, an expanded H-M, space-group symbol showing a more complete representation of the symmetry elements present, and a number that is an index to the order of presentation in the book. This number is useful as an alternative identification of the group and must be given if an orientation other than the standard one is used. (For example, this space group can equally well be described by the symbol *Pmca*, which designates the same set of symmetry elements in different orientations with respect to the coordinate axes.)

Next on the page are two schematic diagrams. The one on the left shows the set of equivalent positions in the unit cell that are generated by the symmetry operations from a *general*, i.e., lacking any symmetry, position with coordinates x, y, and z. A line through the middle of the circle indicates that two positions are superposed in this projection, and a comma indicates that the environment of the point at that position is the mirror image of the environment of the point x, y, z.

The right-hand diagram is a schematic representation of the symmetry elements and their locations in the unit cell, using a system of representation that is illustrated elsewhere in the *International Tables*. In this case, twofold screw axes perpendicular to the page are represented by propeller-like figures, with a hole in the middle showing that a center of symmetry lies on the axis. Twofold rotation axes and twofold screw axes parallel to the page are represented by arrows with two barbs and one barb, respectively. Glide planes perpendicular to the page are shown by dashed and dotted lines: The dashed line indicates a translation parallel to the page, whereas the dotted line indicates a perpendicular translation. Finally, the right angle to the upper right shows a mirror plane parallel to the page.

[3]There must be very few practicing crystallographers who have not had the experience of having a solid-state physicist or chemist present them with a photocopy of a page from the book, asking "What does this *mean*?"

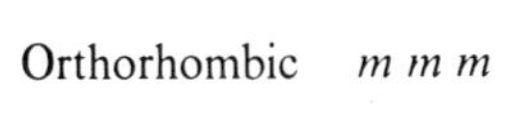

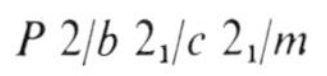

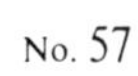

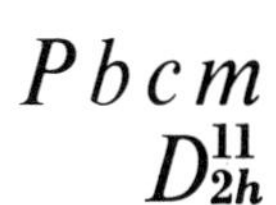

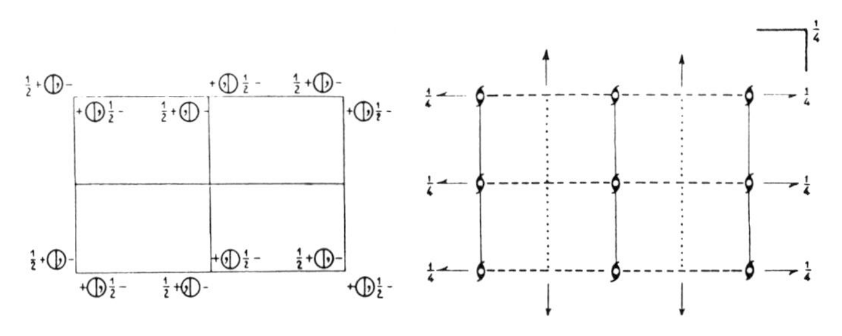

Origin at $\bar{1}$

Number of positions, Wyckoff notation, and point symmetry			Co-ordinates of equivalent positions	Conditions limiting possible reflections
				General:
8	e	1	$x,y,z;\ \bar{x},\bar{y},\tfrac{1}{2}+z;\ x,\tfrac{1}{2}-y,\bar{z};\ \bar{x},\tfrac{1}{2}+y,\tfrac{1}{2}-z;$ $\bar{x},\bar{y},\bar{z};\ x,y,\tfrac{1}{2}-z;\ \bar{x},\tfrac{1}{2}+y,z;\ x,\tfrac{1}{2}-y,\tfrac{1}{2}+z.$	hkl: No conditions $0kl$: $k=2n$ $h0l$: $l=2n$ $hk0$: No conditions $h00$: No conditions $0k0$: $(k=2n)$ $00l$: $(l=2n)$
				Special: as above, plus
4	d	m	$x,y,\tfrac{1}{4};\ \bar{x},\bar{y},\tfrac{3}{4};\ \bar{x},\tfrac{1}{2}+y,\tfrac{1}{4};\ x,\tfrac{1}{2}-y,\tfrac{3}{4}.$	no extra conditions
4	c	2	$x,\tfrac{1}{4},0;\ \bar{x},\tfrac{3}{4},0;\ x,\tfrac{1}{4},\tfrac{1}{2};\ \bar{x},\tfrac{3}{4},\tfrac{1}{2}.$	hkl: $l=2n$
4	b	$\bar{1}$	$\tfrac{1}{2},0,0;\ \tfrac{1}{2},\tfrac{1}{2},0;\ \tfrac{1}{2},0,\tfrac{1}{2};\ \tfrac{1}{2},\tfrac{1}{2},\tfrac{1}{2}.$	hkl: $k=2n$; $l=2n$ (for b and a)
4	a	$\bar{1}$	$0,0,0;\ 0,\tfrac{1}{2},0;\ 0,0,\tfrac{1}{2};\ 0,\tfrac{1}{2},\tfrac{1}{2}.$	

Symmetry of special projections

(001) pgm; $a'=a$, $b'=b$ (100) pgm; $b'=b/2$, $c'=c$ (010) pmm; $c'=c/2$, $a'=a$

Fig. 3.2. Facsimile of a typical space-group page of the *International Tables for X-ray Crystallography*. (Reproduced by permission.)

Below the diagrams is a statement of the point-group symmetry of the surroundings of the origin. Next are the coordinates of the most general set of symmetry-related positions, the *general position*. First is given the multiplicity, the number of equivalent positions in the unit cell. Next is a lowercase italic letter, the Wyckoff symbol,[4] by which position is often identified in a publication, the point-group symmetry of the surroundings of the point, and a list of the coordinates of all equivalent positions.

[4] After R. W. G. Wyckoff, who tabulated them.

On the right are listed a set of diffraction conditions. Diffracted intensities are proportional to $|F|^2$, where

$$F(h,k,l) = \sum_j f_j \exp\left[2\pi i(hx_j + ky_j + lz_j) - B_j\right].$$

Here f_j is an *atomic scattering factor* for an atom located at x_j, y_j, z_j, and B_j is called a *temperature factor* if the speaker was trained as a crystallographer, and a *Debye-Waller factor* if the speaker was trained as a solid-state physicist. The sum is taken over all atoms in the unit cell. If the conditions in the right-hand column are *not* met for a reflection of the given class, $|F| \equiv 0$, and the reflection is said to be *systematically absent*. (The value of $|F|$ may be very small, by chance, for reasons not related to symmetry. In this case the reflection is *accidentally absent*.) Systematic absences are characteristic of the presence of screw axes and glide planes and are, therefore, an important key to the determination of a space group.

If an atomic position lies on a rotation axis, a mirror plane, or a center of symmetry, some of the equivalent general positions become *degenerate*. In our example, if $(x, y, z) = (0,0,0)$, then $(\bar{x}, \bar{y}, \bar{z}) = (0,0,0)$ also, and the points coincide. Such relationships give rise to *special positions*, with multiplicities that are some integral submultiple of the multiplicity of the general position. The possible special positions are listed below the general position, each with its multiplicity, a Wyckoff symbol, the point-group symmetry of its surroundings, and the coordinates of the equivalent positions. Each special position has fewer than three independent-variable parameters. The special values of the coordinates may give rise to additional restrictions on the indices of reflections that must be satisfied if that atom is to contribute to those reflections. These additional conditions are listed on the right for each special position.

Finally, at the bottom of the page, two-dimensional, space-group symbols are given for certain special projections of the unit cell. In the early days of structural crystallography, much work was done in inferring three-dimensional structures from two-dimensional projections, but recently most structural determination work has used full, three-dimensional data, and the projection information is less useful in structural studies than it once was.

Chapter 4

Vectors

We have defined a $1 \times n$ matrix as a row vector and an $n \times 1$ matrix as a column vector. A particularly important application of matrix algebra arises when $n = 3$, and we are dealing with occurrences in the three-dimensional, real space we live in. A vector is then a quantity with a magnitude and a direction. Let $\mathbf{x}$ be the column vector

$$\mathbf{x} = \begin{bmatrix} x_1 \\ x_2 \\ x_3 \end{bmatrix}.$$

We define its magnitude by $|\mathbf{x}| = (x_1^2 + x_2^2 + x_3^2)^{1/2} = (\mathbf{x}^T\mathbf{x})^{1/2}$ and consider a line whose direction cosines, with respect to an orthonormal system, are given by $\cos \alpha_i = x_i/|\mathbf{x}|$. We can reach any point in this three-dimensional space by moving a distance x_1 parallel to axis 1, a distance x_2 parallel to axis 2, and a distance x_3 parallel to axis 3. If we multiply the vector $\mathbf{x}$ by the inversion matrix,

$$\mathbf{R} = \begin{bmatrix} -1 & 0 & 0 \\ 0 & -1 & 0 \\ 0 & 0 & -1 \end{bmatrix},$$

we obtain the vector $-\mathbf{x}$, which corresponds to a displacement in the opposite direction from the origin. If we apply forces to a particle proportional to x_1, x_2, and x_3 parallel to the 1, 2, and 3 axes, respectively, the acceleration will be in the direction of, and proportional to, $\mathbf{x}$. And, again, a force in the opposite direction will produce an acceleration in the opposite direction. Displacements, linear velocities, forces, and accelerations are examples of *polar vectors*, and they are characterized by this property of reversing direction as a result of a space inversion.

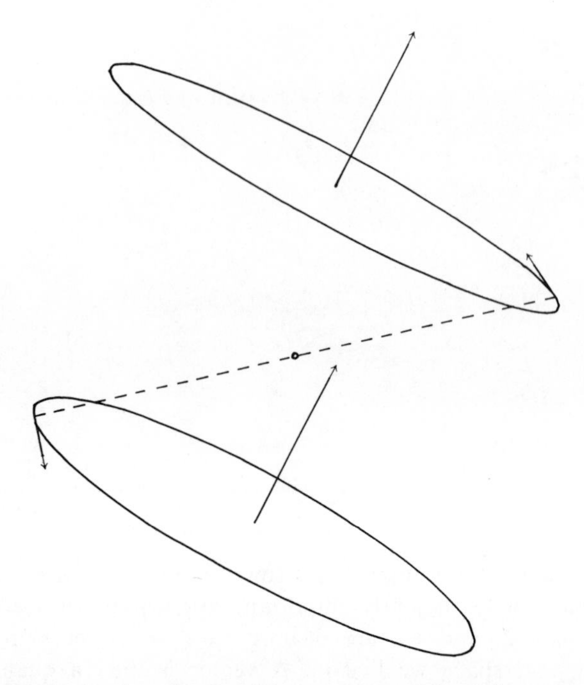

Fig. 4.1. Two spinning hoops related by a center of inversion spin in the same direction. The axial vectors describing their angular velocities therefore also point in the same direction. Axial vectors are invariant under inversion.

Consider, however, a child's hoop. The orientation in space of its plane can be described by the direction cosines of the normal to the plane with respect to an orthonormal coordinate system. If it is spinning, its angular velocity can be specified by a single number $|\mathbf{\Omega}|$, and its angular displacement from an arbitrary starting point by a single number $|\mathbf{\Lambda}|$. The triple of numbers λ_1, λ_2, λ_3, defined by $\lambda_i = |\mathbf{\Lambda}|\cos\alpha_i$, where α_1, α_2, and α_3 are the angles between the normal to the plane and the x, y, and z axes, also forms a vector. We define, somewhat arbitrarily, a rotation in the right-hand screw direction as corresponding to a positive value of the displacement. If we now consider a second hoop related to the first by an inversion through the origin, we discover (see Fig. 4.1) that the second hoop is spinning in the *same direction* as the first; its angular velocity is *invariant under inversion*. Angular displacement and angular velocity—and also torque, angular acceleration, and angular momentum—are examples of *axial vectors*. They are characterized by invariance under space inversion.

Scalar and Vector Products

There are two linear operations with vectors that are so useful and so commonly encountered that they have special names and notations. Consider two column vectors, **a** and **b**. The product $\mathbf{a}^T\mathbf{b} = a_1b_1 + a_2b_2 + a_3b_3$ is a 1×1 matrix called the *scalar product of* **a** *and* **b**. This relationship is often written $\mathbf{a} \cdot \mathbf{b}$. Hence it is often referred to as the "dot product" of **a** and **b**. Further, it is sometimes referred to as the *inner product* of **a** and **b**. Let **a** and **b** be displacement vectors from the origin to two points in an orthonormal coordinate system. The vector $\mathbf{c} = \mathbf{b} - \mathbf{a}$ is the third side of a triangle, two of whose sides are **a** and **b**. By the law of cosines

$$|\mathbf{c}|^2 = |\mathbf{a}|^2 + |\mathbf{b}|^2 - 2|\mathbf{a}||\mathbf{b}|\cos\theta,$$

where θ is the angle included between **a** and **b**. But

$$|\mathbf{c}|^2 = c_1^2 + c_2^2 + c_3^2 = \mathbf{c}^T\mathbf{c},$$

and

$$c_i = b_i - a_i$$

so

$$|\mathbf{c}|^2 = b_1^2 - 2b_1a_1 + a_1^2 + b_2^2 - 2b_2a_2 + a_2^2 + b_3^2 - 2b_3a_3 + a_3^2.$$

Also

$$|\mathbf{c}|^2 = a_1^2 + a_2^2 + a_3^2 + b_1^2 + b_2^2 + b_3^2$$
$$- 2\left(a_1^2 + a_2^2 + a_3^2\right)^{1/2}\left(b_1^2 + b_2^2 + b_3^2\right)^{1/2}\cos\theta.$$

Equating the two expressions for $|\mathbf{c}|^2$, and subtracting like terms from both sides, we obtain

$$a_1b_1 + a_2b_2 + a_3b_3 = \mathbf{a}^T\mathbf{b} = \mathbf{a} \cdot \mathbf{b} = |\mathbf{a}||\mathbf{b}|\cos\theta.$$

All of these expressions are therefore equivalent definitions of the scalar product. If $\mathbf{a} \cdot \mathbf{b} = 0$ when neither $|\mathbf{a}| = 0$ nor $|\mathbf{b}| = 0$, then $\cos\theta = 0$, and the two vectors are at right angles, or *orthogonal.*

The other common relationship between two vectors, **a** and **b**, is the vector

$$\mathbf{c} = \begin{bmatrix} a_2b_3 - a_3b_2 \\ a_3b_1 - a_1b_3 \\ a_1b_2 - a_2b_1 \end{bmatrix} = \begin{bmatrix} 0 & -a_3 & a_2 \\ a_3 & 0 & a_1 \\ -a_2 & a_1 & 0 \end{bmatrix} \begin{bmatrix} b_1 \\ b_2 \\ b_3 \end{bmatrix}.$$

This relationship is usually denoted by $\mathbf{a} \times \mathbf{b}$ and is called the *vector product* or the *cross product* of **a** and **b**. It is sometimes called the *outer product*, although that term is also used for the matrix product $\mathbf{a}\mathbf{b}^T$, which is sometimes called the *direct product*. To avoid further confusion we shall not use the term outer product to refer to either relationship.

It is readily verified that $\mathbf{a} \cdot \mathbf{a} \times \mathbf{b} = \mathbf{b} \cdot \mathbf{a} \times \mathbf{b} = 0$, so that, if $\mathbf{a} \neq 0$, $\mathbf{b} \neq 0$, and $\mathbf{a} \neq \mathbf{b}$, then $\mathbf{a} \times \mathbf{b}$ is a vector that forms right angles with both $\mathbf{a}$ and $\mathbf{b}$. It is also readily verified that $|\mathbf{a} \times \mathbf{b}|^2 = |\mathbf{a}|^2|\mathbf{b}|^2 - (\mathbf{a} \cdot \mathbf{b})^2$, from which it follows that $|\mathbf{a} \times \mathbf{b}| = |\mathbf{a}||\mathbf{b}|\sin\theta$.

A great many useful relationships follow from the fact that $\mathbf{a} \cdot \mathbf{b}$ is equal to the product of the magnitude of $\mathbf{a}$ and the magnitude of the projection of $\mathbf{b}$ upon $\mathbf{a}$. One of these arises if we consider a parallelepiped defined by three vectors, $\mathbf{a}$, $\mathbf{b}$, and $\mathbf{c}$. $|\mathbf{a} \times \mathbf{b}|$ is the area of the parallelogram formed by $\mathbf{a}$ and $\mathbf{b}$, and the altitude is the projection of $\mathbf{c}$ upon $\mathbf{a} \times \mathbf{b}$. Therefore $(\mathbf{a} \times \mathbf{b}) \cdot \mathbf{c}$ is the volume of the parallelepiped. Expansion of the definitions shows that

$$(\mathbf{a} \times \mathbf{b}) \cdot \mathbf{c} = (\mathbf{c} \times \mathbf{a}) \cdot \mathbf{b} = (\mathbf{b} \times \mathbf{c}) \cdot \mathbf{a} = \begin{vmatrix} a_1 & a_2 & a_3 \\ b_1 & b_2 & b_3 \\ c_1 & c_2 & c_3 \end{vmatrix}.$$

In other words, the volume of a parallelepiped formed by three noncoplanar vectors is the value of the determinant composed of the components of the three vectors with respect to an orthonormal coordinate system.

The Reciprocal Lattice

University physics courses in elementary physics usually give Bragg's law for the diffraction of plane waves by a crystal in the form

$$n\lambda = 2d\sin\theta,$$

where λ is the wavelength of the plane waves, θ is the glancing angle of incidence of the wave normal on crystal planes, d is the spacing between successive planes of the crystal lattice, and n is an integer. Because all properties of a crystal must remain invariant under a lattice translation, it follows that a translation from one lattice point to an adjacent one must cross an integral number of identical planes (unless the crystal plane we are considering is parallel to the lattice planes, in which case the integral number is zero). The plane nearest to the origin is defined by the three points at the ends of the vectors $\mathbf{a}/h$, $\mathbf{b}/k$, and $\mathbf{c}/l$, where h, k, and l are integers, provided none of them is zero. If one or two of them are zero (all three cannot be), the plane is parallel to the corresponding lattice vector and the plane is still uniquely defined. The integers h, k, and l are the *Miller indices* defining the plane. (We shall frequently denote the column vector whose elements are h, k, and l by $\mathbf{h}$.)

The lattice spacing, d, is the distance, along a line passing through the origin and perpendicular to this first plane, from the origin to the point of intersection with the plane. The magnitude of d can be expressed in terms of $|\mathbf{a}|$, $|\mathbf{b}|$, and $|\mathbf{c}|$, but the expressions, except in the cases of highly symmetric lattices, are very complex. They can be made much simpler if they are expressed in terms of the *reciprocal lattice*, which is defined by three new, noncoplanar vectors, $\mathbf{a}^*$, $\mathbf{b}^*$, and $\mathbf{c}^*$, such that

$$\mathbf{a}^* = (\mathbf{b} \times \mathbf{c})/V,$$

$$\mathbf{b}^* = (\mathbf{c} \times \mathbf{a})/V, \quad \text{and}$$

$$\mathbf{c}^* = (\mathbf{a} \times \mathbf{b})/V,$$

where $V = \mathbf{a} \cdot (\mathbf{b} \times \mathbf{c})$ is the volume of the real-space unit cell. Inspection shows that the reciprocal lattice vectors have the following properties:

$$\mathbf{a} \cdot \mathbf{a}^* = \mathbf{b} \cdot \mathbf{b}^* = \mathbf{c} \cdot \mathbf{c}^* = 1, \quad \text{and}$$

$$\mathbf{a} \cdot \mathbf{b}^* = \mathbf{a} \cdot \mathbf{c}^* = \mathbf{b} \cdot \mathbf{a}^* = \mathbf{b} \cdot \mathbf{c}^* = \mathbf{c} \cdot \mathbf{a}^* = \mathbf{c} \cdot \mathbf{b}^* = 0.$$

Let us assume that the Miller indices are all different from zero, so that a plane is defined by the three points $\mathbf{a}/h$, $\mathbf{b}/k$, and $\mathbf{c}/l$. We form the vector $\mathbf{d}^* = h\mathbf{a}^* + k\mathbf{b}^* + l\mathbf{c}^*$. The vector $\mathbf{x} = \mathbf{a}/h - \mathbf{b}/k$ is a vector lying in the plane. The scalar product $\mathbf{x} \cdot \mathbf{d}^* = \mathbf{a} \cdot \mathbf{a}^* - \mathbf{b} \cdot \mathbf{b}^* = 0$. Similarly, $(\mathbf{a}/h - \mathbf{c}/l) \cdot \mathbf{d}^* = 0$, whereas $\mathbf{a} \cdot \mathbf{d}^* = \mathbf{b} \cdot \mathbf{d}^* = \mathbf{c} \cdot \mathbf{d}^* = 1$. The first two relationships show that the vector $\mathbf{d}^*$ is perpendicular to the plane whose indices are h, k, and l. The value of d is therefore the projection of $\mathbf{a}$, $\mathbf{b}$, or $\mathbf{c}$ on $\mathbf{d}^*$. That being true, the third relationship shows that $d = 1/|\mathbf{d}^*|$. If one or two of h, k, or l are zero, it can easily be shown that these properties of $\mathbf{d}^*$ are still satisfied.

Looking at Figure 4.2, consider a vector, $\mathbf{s}_0$, in the propagation direction of a plane wave, with magnitude $1/\lambda$, and another vector, $\mathbf{s}_f$, in the

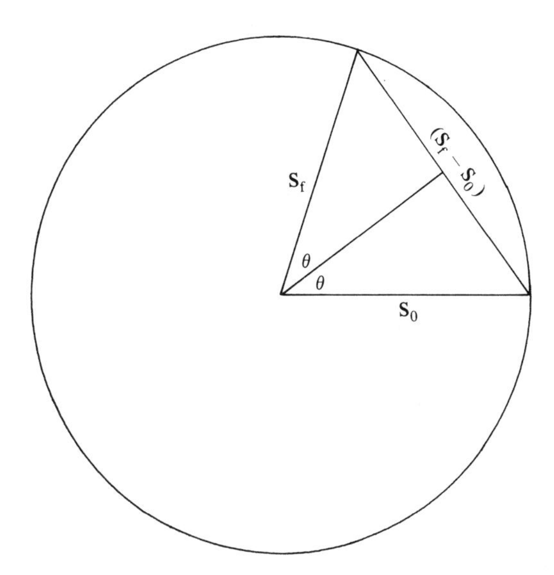

Fig. 4.2. Ewald's construction for solution of the Bragg equation. If $(\mathbf{s}_f - \mathbf{s}_0)$ is a vector of the reciprocal lattice, the Bragg equation is satisfied for the corresponding set of planes.

propagation direction of a reflected plane wave, also with magnitude $1/\lambda$. The difference $(\mathbf{s}_f - \mathbf{s}_0)$ is a vector perpendicular to the bisector of $\mathbf{s}_f$ and $\mathbf{s}_0$, and its magnitude is $|\mathbf{s}_f - \mathbf{s}_0| = 2|\mathbf{s}|\sin\theta = 2\sin\theta/\lambda$. The conditions for Bragg's law to be satisfied can therefore be written as the vector equation $\mathbf{d}^* = \mathbf{s}_f - \mathbf{s}_0$. The integer, n, in the earlier expression for Bragg's law is usually incorporated in $\mathbf{d}^*$, where it appears as a common divisor of the Miller indices.

The terminal points of all possible vectors, $\mathbf{s}_f$, that correspond to possible diffraction peaks lie on the surface of a sphere passing through the origin of the reciprocal lattice, with a radius $1/\lambda$. If any node of the reciprocal lattice lies on the surface of this sphere, therefore, Bragg's law will be satisfied for the family of lattice planes corresponding to that node. This construction was first introduced by P. P. Ewald. For this reason the sphere in reciprocal space is often called the *Ewald sphere*.

The Orientation Matrix

We saw earlier (page 19) that the length, in real-space units, of a vector, $\mathbf{x}$, expressed in crystal lattice fractional units, could be written in the form $|\mathbf{x}| = (\mathbf{x}^T\mathbf{G}\mathbf{x})^{1/2}$, where $\mathbf{G}$ is the real-space metric tensor,

$$\mathbf{G} = \begin{pmatrix} \mathbf{a}\cdot\mathbf{a} & \mathbf{a}\cdot\mathbf{b} & \mathbf{a}\cdot\mathbf{c} \\ \mathbf{a}\cdot\mathbf{b} & \mathbf{b}\cdot\mathbf{b} & \mathbf{b}\cdot\mathbf{c} \\ \mathbf{a}\cdot\mathbf{c} & \mathbf{b}\cdot\mathbf{c} & \mathbf{c}\cdot\mathbf{c} \end{pmatrix}.$$

We saw, further, that the vector itself could be transformed into an orthonormal coordinate system by a linear transformation of the form $\mathbf{x}' = \mathbf{A}\mathbf{x}$, where $\mathbf{A}$ is an upper triangular matrix such that $\mathbf{A}^T\mathbf{A} = \mathbf{G}$. We can perform similar operations on the reciprocal lattice by using the reciprocal space metric tensor

$$\mathbf{G}^{-1} = \begin{pmatrix} \mathbf{a}^*\cdot\mathbf{a}^* & \mathbf{a}^*\cdot\mathbf{b}^* & \mathbf{a}^*\cdot\mathbf{c}^* \\ \mathbf{a}^*\cdot\mathbf{b}^* & \mathbf{b}^*\cdot\mathbf{b}^* & \mathbf{b}^*\cdot\mathbf{c}^* \\ \mathbf{a}^*\cdot\mathbf{c}^* & \mathbf{b}^*\cdot\mathbf{c}^* & \mathbf{c}^*\cdot\mathbf{c}^* \end{pmatrix}.$$

We can choose an upper triangular matrix, $\mathbf{B}$, such that $\mathbf{B}^T\mathbf{B} = \mathbf{G}^{-1}$, which will transform a reciprocal lattice vector into an orthonormal coordinate system with x parallel to $\mathbf{a}^*$, y perpendicular to x and lying in the $\mathbf{a}^*\mathbf{b}^*$ plane, and z perpendicular to x and y and forming a right-handed system.

There is, however, an infinite number of ways to choose a matrix $\mathbf{B}$ such that $\mathbf{B}^T\mathbf{B} = \mathbf{G}^{-1}$, and the upper triangular one is not necessarily the best one. In the systematic scanning through reciprocal space for the collection of diffraction data it is convenient to choose the orthonormal coordinate system so that x is parallel to $\mathbf{a}^*$, $\mathbf{b}^*$, or $\mathbf{c}^*$ depending on whether h, k, or l is the index most frequently changed, and z parallel to $\mathbf{a}$, $\mathbf{b}$, or $\mathbf{c}$ depending on whether h, k, or l is the index least frequently changed. This choice leads to the simplest procedure for determining the ranges of the various indices for lattice nodes lying within the limiting sphere defined by the maximum value of $2\sin\theta/\lambda$. A practical way to find the appropriate form of **B** is to rearrange the rows and columns of $\mathbf{G}^{-1}$ so that the reciprocal vector corresponding to the most frequently changed index is in position 11 and the vector corresponding to the least frequently changed index is in position 33, and then perform a Cholesky decomposition. It is also convenient to multiply each nonzero element of **B** by $\lambda/2$, leading to the particularly simple relationship $\sin\theta = |\mathbf{Bh}|$ for the satisfaction of the Bragg conditions.

The product **Bh** gives a vector normal to a family of crystal lattice planes, referred to an orthonormal coordinate system attached to the crystal. To find a Bragg reflection from those planes we need also an orthogonal rotation matrix that relates the coordinate system attached to the crystal to one attached to the diffractometer. There are a number of ways to determine this matrix. One of the simplest is as follows: First, find two reflections and carefully measure the values of the Eulerian angles ϕ, χ, and ω for which the reflected intensity is a maximum. The value of $\mathbf{d}^*$, determined from the observed value of the Bragg angle, θ, and the value of the included angle between the normals, determined from the Eulerian angles, should give us a reasonable assurance that we know the correct indices for these reflections. Denote the indices of the two reflections by $\mathbf{h}_1$ and $\mathbf{h}_2$. Now define an orthonormal coordinate system with the x axis parallel to $\mathbf{h}_1$ and the z axis parallel to $\mathbf{h}_1 \times \mathbf{h}_2$. The direction cosines of $\mathbf{h}_1$ and $\mathbf{h}_2$, and therefore the direction cosines of the three axes of the orthonormal system, can be determined, relative to the system fixed with respect to the crystal, by the matrix products $\mathbf{Bh}_1$ and $\mathbf{Bh}_2$. Denote the transformation from the crystal system to the reflection system by $\mathbf{U}_c$. The direction cosines of $\mathbf{h}_1$ and $\mathbf{h}_2$ relative to the system fixed with respect to the diffractometer are specified by the Eulerian angles, so that we may determine the matrix, $\mathbf{U}_d$, which transforms from the reflection system to the diffractometer system. The transformation from the crystal system to the diffractometer system is then the product $\mathbf{U} = \mathbf{U}_d\mathbf{U}_c$. Therefore, the components of the vector $\mathbf{d}^*$ relative to the diffractometer system, when the diffractometer angles are all set to zero, are given by the product $\mathbf{x} = \mathbf{UBh}$. The matrix **UB**, which is the product of a matrix, **B**, dependent on the dimensions of the unit cell of the crystal, and a matrix, **U**, which is an orthogonal rotation matrix relating an orthonormal coordinate system fixed with respect to the crystal to another orthonormal coordinate system fixed with respect to the diffractometer, is called the *orientation matrix*.

If we specify the diffractometer system as one with the z axis parallel to the 2θ axis of the diffractometer and the x axis parallel to the direction of the incident X-ray or neutron beam, and we further specify that the χ axis is parallel to x when $\omega = 0$, then $\mathbf{d}^* = \mathbf{UBh}$ is brought into the position for Bragg reflection by the angles $\chi = \arcsin(d_z^*/|\mathbf{d}^*|)$, $\phi = \arctan(d_x^*/d_y^*)$, and $\omega = \theta = \arcsin|\mathbf{d}^*|$. These angles specify what is commonly known as the "bisecting position" for a four-circle diffractometer. There is an infinite number of other possible combinations of Eulerian angles that satisfy the

Bragg conditions. They have practical uses, but discussion of them is beyond the scope of this book.

Zones and Forms

If the normals to the lattice planes in a crystal are projected onto the surface of a sphere, it can be observed that there are many families of planes for which the projected points fall on great circles. These families of planes are known as *zones*, and the common direction to which the normals are perpendicular and the planes are parallel is a *zone axis*. Given two planes whose normals are the vectors $\mathbf{d}_1^*$ and $\mathbf{d}_2^*$, their zone axis is $\mathbf{d}_1^* \times \mathbf{d}_2^*$. As is the case for the expression for the value of d, a substantial simplification results if we make use of the relationships between the direct and reciprocal lattices. It can easily be shown that

$$(\mathbf{d}_1^* \times \mathbf{d}_2^*) = u\mathbf{a} + v\mathbf{b} + w\mathbf{c},$$

where $u = k_1l_2 - k_2l_1$, $v = l_1h_2 - l_2h_1$, and $w = h_1k_2 - h_2k_1$. If we designate the vectors (u, v, w), (h_1, k_1, l_1), and (h_2, k_2, l_2) by $\mathbf{u}$, $\mathbf{h}_1$, and $\mathbf{h}_2$, respectively, this relationship can be written $\mathbf{u} = \mathbf{h}_1 \times \mathbf{h}_2$. The relationship applies without regard to the symmetry (or lack thereof) of the lattice, so long as $\mathbf{u}$ is expressed in terms of the direct lattice vectors and $\mathbf{h}_1$ and $\mathbf{h}_2$ are expressed in terms of the reciprocal lattice vectors.

If a set of indices denotes a particular family of parallel planes, they are written in parentheses, e.g., (hkl). The indices of a zone axis are enclosed in square brackets, e.g., $[uvw]$. A family of planes related by the operations of a point group is known as a *form* and is designated by enclosing the indices in curly braces e.g., $\{hkl\}$. Finally, a family of zone axes that are related by symmetry is designated by enclosing the indices in angle brackets, e.g., $\langle uvw \rangle$. If a natural crystal is enclosed by planes belonging to a form (or a small number of forms) it is said to exhibit a *habit*. For example, a cubic crystal enclosed by the faces of a rhombic dodecahedron is said to have a $\{110\}$ habit.

Sublattices and Superlattices

Consider a lattice that is defined by a set of three noncoplanar vectors, $\mathbf{a}$, $\mathbf{b}$, and $\mathbf{c}$. Let us define three other vectors, $\mathbf{a}'$, $\mathbf{b}'$, and $\mathbf{c}'$, by the relationships

$$\mathbf{a}' = S_{11}\mathbf{a} + S_{12}\mathbf{b} + S_{13}\mathbf{c},$$

$$\mathbf{b}' = S_{21}\mathbf{a} + S_{22}\mathbf{b} + S_{23}\mathbf{c},$$

$$\mathbf{c}' = S_{31}\mathbf{a} + S_{32}\mathbf{b} + S_{33}\mathbf{c},$$

where all of the elements of the matrix, $\mathbf{S}$, are integers, and the determinant, $|\mathbf{S}|$, is an integer, n, > 1. The vectors $\mathbf{a}'$, $\mathbf{b}'$, and $\mathbf{c}'$ define a new lattice whose nodes are all nodes of the original lattice, but whose unit cell volume is n times the volume of the cell of the original lattice. Such a lattice is known as a *superlattice* of the original lattice. The lattice defined by **a**, **b**, and **c** is, correspondingly, a *sublattice* of the lattice defined by **a**′, **b**′, and **c**′. If a lattice, L_2, is a superlattice of another lattice, L_1, then the reciprocal lattice, L_2^*, is a sublattice of the lattice, L_1^*, and vice versa.

For any integer, n, there is a finite number of matrices, **S**, that produce distinct superlattices.[1] Although the number quickly gets large as n gets large, the number for $n \leqslant 4$ is small enough so that they may be listed in a table. The matrices for $n = 2$, 3, and 4, are given in Appendix C. Many crystallographic phase transitions take place between two structures for which one lattice is a superlattice of the other. A knowledge of all of the possible superlattices of a given lattice makes it possible to examine an inclusive set of possible structures of a new phase. This is particularly useful if, as is frequently the case, one or both phases are available for study only as polycrystalline samples. A fact that should be carefully noted is that there may be more than one distinct superlattice that gives peaks in a powder pattern in identical positions. The multiplicities are different, but it is necessary to pay careful attention to intensities to distinguish one from another.

[1]See A. Santoro and A. D. Mighell, Properties of crystal lattices: The derivative lattices and their determination. Acta Cryst. A28: 284–287, 1972.

Chapter 5

Tensors

As we saw in Chapter 1, a matrix is an array of numbers that can be used to express a linear relationship between one set of quantities and another. A matrix is thus a very general mathematical concept. If, however, the matrix is used in the physical world to express linear relationships between measurable quantities, such as those between "causes" and "effects," we are dealing with a restricted type of matrix called a *tensor*. Suppose, for example, that we apply an electric field, a vector quantity, to a crystal and measure the electric polarization, another vector quantity. The polarization, $\mathbf{P}$, can be expressed as a function of the field, $\mathbf{E}$, by $\mathbf{P} = \mathbf{AE}$. It is obvious that it makes no difference what system of coordinates is used to express the components of $\mathbf{P}$ and $\mathbf{E}$, and that the component of $\mathbf{P}$ along a direction fixed in space, parallel to a vector, $\mathbf{v}$, for example, is the same in any coordinate system. We can express this statement mathematically by saying that $\mathbf{v} \cdot \mathbf{P} = \mathbf{v} \cdot (\mathbf{AE}) = \mathbf{v}^T\mathbf{AE} = C$, a constant independent of coordinate systems. Or, in other words, the quantity $\mathbf{v} \cdot (\mathbf{AE})$ is invariant under all rotations of the coordinate axes.

Consider an orthogonal rotation matrix, $\mathbf{R}$, that transforms the coordinates in such a way that $\mathbf{v}' = \mathbf{Rv}$, and $\mathbf{E}' = \mathbf{RE}$. The invariance condition says that $\mathbf{v}' \cdot (\mathbf{A}'\mathbf{E}') = \mathbf{v} \cdot (\mathbf{AE}) = C$, or $(\mathbf{Rv})^T\mathbf{A}'(\mathbf{RE}) = \mathbf{v}^T\mathbf{AE}$. This becomes $\mathbf{v}^T\mathbf{R}^T\mathbf{A}'\mathbf{RE} = \mathbf{v}^T\mathbf{AE}$. The equation is satisfied if $\mathbf{A}' = (\mathbf{R}^T)^{-1}\mathbf{AR}^{-1}$, or, since $\mathbf{R}$ is orthogonal, $\mathbf{A}' = \mathbf{RAR}^T$. (It should be noted that the first of these expressions is true whether or not $\mathbf{R}$ is orthogonal.)

The matrix $\mathbf{A}$ relating the vector $\mathbf{P}$ and the vector $\mathbf{E}$ is a *tensor of the second rank*. It is defined by stating the transformation property, which may be stated explicitly as follows: If a vector, $\mathbf{x}$, is transformed from one coordinate system to another by $\mathbf{x}' = \mathbf{Rx}$, if $\mathbf{Q} = \mathbf{R}^{-1}$, and

$$A'_{ij} = \sum_{k=1}^{n} \sum_{l=1}^{n} Q_{ik} Q_{jl} A_{kl},$$

then **A** is a tensor of the second rank. Because **A** is a particular type of square matrix, it has other properties of square matrices. Specifically, it has eigenvalues and eigenvectors, so that there will be some directions, **u**, in which $\mathbf{P} = \mathbf{AE} = \lambda\mathbf{E}$, where λ is an eigenvalue. Since we are referring to measurable physical quantities, the eigenvalues must be real numbers and, further, because a diagonal matrix is symmetric, and a matrix transformation of the form of the tensor transformation keeps a symmetric matrix symmetric, most, if not all, second-rank tensors with physical significance are symmetric.

Covariance and Contravariance

Albert Einstein, in his formulation of the metric cosmologies that became the general theory of relativity, introduced the concept of two types of tensors, which he named covariant and contravariant.[1] A vector is covariant if its elements increase (numerically) in size when the units in which the coordinate system is measured increase in size. The gradient of a physical quantity is covariant. For example, if the gradient of electron density is a quantity x per angstrom, it will be $10x$ per nanometer. A vector is contravariant if its elements decrease (numerically) in size when the units in which the coordinate system is measured increase in size. An interatomic vector is contravariant—if a distance is x when the units are angstroms, it will be $.1x$ if the units are nanometers. A second-rank tensor may be formed by the direct product, $\mathbf{uv}^T$, of two vectors. As the vectors are covariant or contravariant, the resulting tensor may be covariant or contravariant with respect to either or both of its indices. Contravariant tensors are often designated by giving their indices as superscripts, as in A^{ij}, keeping the subscripts, A_{ij}, for covariant tensors, and a mixture, A_j^i, for mixed type. This convention is honored far more in the breach than the observance, however, so the position of the indices should never be relied upon to indicate the nature of the tensor.

We have discussed earlier the direct-space and reciprocal-space metric tensors. The direct-space metric tensor is contravariant, and is therefore sometimes called the *contravariant metric tensor*. The reciprocal-space metric tensor is covariant (with reference to the units of measure in direct space), and is sometimes called the *covariant metric tensor*. As we shall see,

[1] If Einstein's intention, when he introduced these terms, was to clarify something, he certainly did not succeed. I often wonder if they were the inspiration of the second of the following heroic couplets—the first is Alexander Pope's epitaph for Sir Isaac Newton—attributed by the *Oxford Dictionary of Quotations* to Sir John Collings Squire.

> Nature and nature's laws lay hid in night.
> God said, "Let Newton be!" and all was light.
> It did not last. The devil, howling "Ho!
> Let Einstein be!" restored the status quo.

anisotropic temperature factors are the elements of a second-rank tensor that can take a variety of forms, depending on the units in which they are expressed. This tensor is contravariant.

The Multivariate Normal Distribution

If $\mathbf{x}$ is an n-dimensional column vector, the function

$$N(\mathbf{x}) = (2\pi)^{-n/2}|\boldsymbol{\Sigma}|^{-1/2}\exp\left[-(1/2)\mathbf{x}^T\boldsymbol{\Sigma}^{-1}\mathbf{x}\right]$$

is the multivariate normal distribution in n dimensions. $\boldsymbol{\Sigma}$ denotes an n-dimensional, second-rank, symmetric tensor known as the *variance-covariance matrix*. If n is equal to one, the function reduces to the well-known Gaussian distribution function

$$G(x,\sigma) = (2\pi\sigma^2)^{-1/2}\exp\left[-(1/2)(x^2/\sigma^2)\right]$$

where σ^2 denotes the *variance* of x. If n is greater than one, and

$$\int_{x_0}^{x_0+\Delta X} N(\mathbf{x})\,d\mathbf{x}$$

is the probability that a random vector, $\mathbf{x}$, will be found to have a value such that $x_{0i} \leqslant x_i \leqslant x_{0i} + \Delta x_i$ for all i, then $N(\mathbf{x})$ is the *joint probability density function* for the quantities x_i. The factor multiplying the exponential is a *normalizing factor*, so that the integral

$$\int_{-\infty}^{+\infty} N(\mathbf{x})\,d\mathbf{x} = 1$$

when the integral is taken over all variables. The function

$$M(x_i) = \int_{-\infty}^{+\infty} N(\mathbf{x})\,d\mathbf{x},$$

where the integration is performed over all components of $\mathbf{x}$ *except* x_i, is the *marginal probability density function*, or *marginal distribution function* for the variable x_i. It is the probability of finding a value of x_i between x_{0i} and $x_{0i} + \Delta x_i$ irrespective of the values of all the other variables. If all other components of $\mathbf{x}$ are given particular values, e.g., $x_j = c_j$ for all $j \neq i$ then the density function (renormalized so that the total probability is still equal to one) is the *conditional probability density function*, or the *conditional distribution function* of x_i for $x_j = c_j$.

$\boldsymbol{\Sigma}$ and $\boldsymbol{\Sigma}^{-1}$ are real, positive definite matrices, and therefore have real, positive eigenvalues. It follows that there is an orthogonal transformation, $\mathbf{R}$, such that $\mathbf{R}\boldsymbol{\Sigma}\mathbf{R}^T$ (or $\mathbf{R}\boldsymbol{\Sigma}^{-1}\mathbf{R}^T$) is diagonal. Denoting $\mathbf{R}\mathbf{x}$ by $\mathbf{x}'$ and

$\mathbf{R\Sigma^{-1}R}$ by $\mathbf{W}$, we have

$$N(x') = (2\pi)^{-n/2}|\mathbf{\Sigma}|^{-1/2}\exp\left[-(1/2)\left(\sum_{i=1}^{n} W_{ii}x_i'^2\right)\right],$$

or

$$N(\mathbf{x}') = \prod_{i=1}^{n}(2\pi W_{ii})^{-1/2}\exp\left[-(1/2)W_{ii}x_i'^2\right].$$

If the joint distribution function of two or more variables, x_i, x_j, etc., is equal to the product of their marginal distribution functions, or equivalently, the marginal distribution function for one variable is equal to its conditional distribution function for all values of the other variables, the variables are said to be *statistically independent*, or *uncorrelated*. If the matrix $\mathbf{\Sigma}$ is not diagonal, if $\Sigma_{ij} \neq 0$ when $i \neq j$, then x_i and x_j are correlated, and the matrix element Σ_{ij} is their *covariance*.

It can be shown[2] that a diagonal element of $\mathbf{\Sigma}, \Sigma_{ii}$, is the variance of the marginal distribution function for the variable x_i. For two variables this can be demonstrated easily as follows:

$$M(x_1) = (2\pi)^{-1}\left(\Sigma_{11}\Sigma_{22} - \Sigma_{12}^2\right)^{-1/2}$$
$$\times \int_{-\infty}^{+\infty}\exp\left[-(1/2)(W_{11}x_1^2 + 2W_{12}x_1x_2 + W_{22}x_2^2)\,dx_2,\right.$$

where $\mathbf{W} = \mathbf{\Sigma}^{-1}$. Setting $C = (2\pi)^{-1}|\mathbf{\Sigma}|^{-1/2}$, and taking factors that do not depend on x_2 outside of the integral, we obtain

$$M(x_1) = C\exp\left[-(1/2)W_{11}x_1^2\right]\int_{-\infty}^{+\infty}\exp\left[-(1/2)(2W_{12}x_1x_2 + W_{22}x_2^2)\right]dx.$$

Completing the square in the argument,

$$M(x_1) = C\exp\left[-(1/2)(W_{11} - W_{12}^2/W_{22})x_1^2\right]$$
$$\times \int_{-\infty}^{+\infty}\exp\left\{-(1/2)\left[(W_{12}/W_{22}^{1/2})x_1 + W_{22}^{1/2}x_2\right]^2\right\}dx_2.$$

Setting $z = [(W_{12}/W_{22}^{1/2})x_1 + W_{22}^{1/2}x_2]$, we obtain $dx_2 = W_{22}^{1/2}dz$, so that

$$M(x_1) = CW_{22}^{1/2}\exp\left[-(1/2)(W_{11} - W_{12}^2/W_2)x_1^2\right]\int_{-\infty}^{+\infty}\exp\left[-(1/2)z^2\right]dz.$$

[2]For a proof of the general, n-dimensional case see W. C. Hamilton, Statistics in Physical Science. The Ronald Press Company, New York, 1964, pp. 129–130.

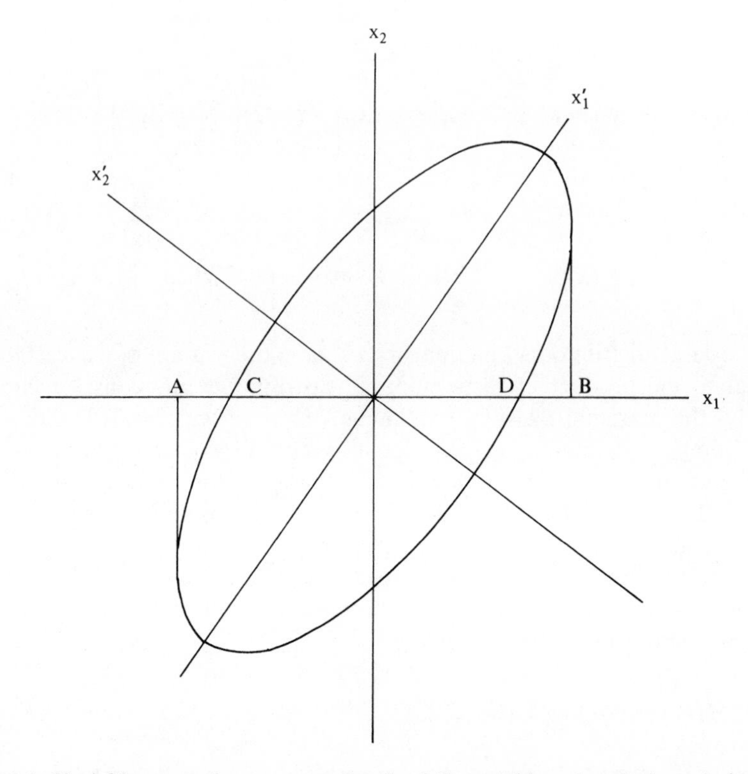

Fig. 5.1. Variables x_1 and x_2 are correlated, while variables x'_1 and x'_2 are independent.

We show in Appendix D that this integral is equal to $(2\pi)^{1/2}$. Now $\Sigma_{11} = W_{22}/(W_{11}W_{22} - W_{12}^2)$, and $W_{22}^{1/2} = [\Sigma_{11}/(\Sigma_{11}\Sigma_{22} - \Sigma_{12}^2)]^{1/2}$. Substitution gives $M(x_1) = (2\pi\Sigma_{11})^{-1/2}\exp[-(1/2)x_1^2/\Sigma_{11}]$, a Gaussian distribution function with variance Σ_{11}, which was to be proved.

Figure 5.1 illustrates independence and correlation graphically. The ellipse is the one *standard deviation* contour for the joint distribution function of two variables, x_1 and x_2. (The standard deviation is the square root of the variance.) The distance $A - B$ is the standard deviation of the marginal distribution of x_1, whereas the shorter segment, $C - D$, is the standard deviation of the conditional distribution of x_1 given $x_2 = 0$. x_1 and x_2 are linear combinations of two statistically independent variables, one of which, x'_2, has a small variance, while the other, x'_1, has a large variance.

Anisotropic Temperature Factors

The *structure amplitude*, or *structure factor*, for a Bragg reflection whose indices are h, k, and l is conventionally given by an expression of the form

$$F_{hkl} = \sum_j f_j \exp\left[2\pi i(hx_j + ky_j + lz_j)\right]\exp\left[-2\pi^2\langle(u_j/d)^2\rangle\right].$$

In this expression f_j is the *atomic scattering factor*, which is a Fourier transform of the distribution of scattering matter in the atom, evaluated at the point in transform space whose coordinates are (h, k, l). For X-rays and for the magnetic scattering of neutrons this is a transform of an electron density distribution. For nonmagnetic scattering of neutrons the scattering matter is concentrated in a nucleus that has dimensions that are very small in comparison with the de Broglie wavelength of the scattered neutrons, and the transform may be treated, for all practical purposes, as a constant. x_j, y_j, and z_j are the coordinates of the equilibrium position of atom j, expressed as fractions of the unit cell edges when the indices, h, k, and l are expressed in reciprocal lattice coordinates.

The second exponential factor in the expression for F is the temperature factor. The expression $\langle (u_j/d)^2 \rangle$ designates the *expected value* of the mean square displacement of the jth atom, from its equilibrium position, in a direction perpendicular to the reflecting planes, given as a fraction of the distance between like planes. The expected value of a quantity, x, [which is also frequently designated by $E(x)$ or $\mathcal{E}(x)$] is defined as

$$\langle x \rangle = \int_{-\infty}^{+\infty} x f(x)\, dx,$$

where $f(x)$ is the probability density function for x, normalized so that

$$\int_{-\infty}^{+\infty} f(x)\, dx = 1.$$

The sum in the expression for F is, strictly, taken over the entire crystal. Because of the periodicity of the crystal lattice, however, the structure factor has nonzero values only at reciprocal lattice points, corresponding to integral values of h, k, and l, and the first exponential factor has the same value for every unit cell. Evaluation of the temperature factor requires an expected value of the mean square displacements of all atoms in the entire crystal at a particular time, a quantity known as the *ensemble average*. Fortunately, the ergodic theorem tells us that the ensemble average of the mean square displacements of a large number of identical particles is equal to the *time average* of the displacement of one particle. The time average of a quantity, x (which is also designated $\langle x \rangle$) is defined by

$$\langle x \rangle = \lim_{T \to \infty} \left[1/(T - T_0) \right] \int_{T_0}^{T} x(t)\, dt.$$

(We shall, hereafter, use $\langle x \rangle$ to designate expected values, whether they be ensemble averages, time averages, or statistical moments.) If the particle is in a harmonic potential well, meaning that is potential energy is a quadratic function of the components of a displacement vector, a quantum mechanical result known as Bloch's theorem tells us that the probability density function for finding a given displacement at any time, T, is

Gaussian. (The harmonic oscillator in quantum mechanics, including Bloch's theorem, is discussed in some detail in Appendix E.)

The temperature factor can be written in the form $\exp\{-2\pi^2\langle(\mathbf{u}\cdot\mathbf{d}^*)^2\rangle\}$, which, expanded, becomes

$$\exp\{-2\pi^2\langle h^2a^{*2}u_x^2 + k^2b^{*2}u_y^2 + l^2c^{*2}u_z^2 + 2hka^*b^*u_xu_y$$

$$+2hla^*c^*u_xu_z + 2klb^*c^*u_yu_z\rangle\},$$

or, in matrix form, $\exp\{-2\pi^2\mathbf{d}^{*T}\mathbf{U}\mathbf{d}^*\}$, where $U_{ij} = \langle u_iu_j\rangle$. **U** is a second-rank tensor and is one of several forms of the temperature factor commonly takes in the literature. Two other forms appear commonly enough so that the relationships among them must be made clear. One, commonly designated **B**, is related to **U** by $B_{ij} = 8\pi^2U_{ij}$. The temperature factor then becomes $\exp\{-(1/4)\mathbf{d}^{*T}\mathbf{B}\mathbf{d}^*\}$. The other, commonly designated by $\boldsymbol{\beta}$, is related to **B** by $B_{ij} = 4\beta_{ij}/a_i^*a_j^*$ (where $a_1^* = |\mathbf{a}^*|, a_2^* = |\mathbf{b}^*|$, and $a_3^* = |\mathbf{c}^*|$). Using this form, the temperature factor is the particularly convenient expression $\exp\{-\mathbf{h}^T\boldsymbol{\beta}\mathbf{h}\}$, where **h** represents **d*** expressed in reciprocal lattice units. Because of the simplicity of this expression, the anisotropic temperature factor tensor usually takes the $\boldsymbol{\beta}$ form in least squares refinement programs.

The expression $\exp\{-\mathbf{x}^T\mathbf{U}\mathbf{x}\}$ is a probability density function that describes the relative probability of finding the nucleus of an atom in an infinitesimal box centered on the point displaced by the vector **x** from the atom's equilibrium position. The equation $\mathbf{x}^T\mathbf{U}\mathbf{x} = C$ describes an ellipsoidal surface on which the probability density is a constant. Such surfaces are called *thermal ellipsoids*. If the crystal lattice is nonorthogonal (a large proportion of all known crystal structures is monoclinic), all three forms of the anisotropic temperature factor tensor describe the ellipsoids in nonorthogonal coordinate systems. To visualize the shapes and orientations of the ellipsoids it is convenient to transform the tensor into an orthonormal coordinate system, which may be done by the matrix transformation[3] $\mathbf{U}' = \mathbf{A}\boldsymbol{\beta}\mathbf{A}^T/2\pi^2$, where **A** is the upper triangular square root (Cholesky decomposition) of the real-space metric tensor, **G**, so that $\mathbf{A}^T\mathbf{A} = \mathbf{G}$.

The eigenvalues and eigenvectors of the matrix, $\mathbf{U}'$ are the *principal axes* of the thermal ellipsoids. In Appendix A we discuss the computation of the eigenvalues of a general, symmetric 3×3 matrix, and their derivatives with respect to the matrix elements, which will enable us to determine the standard deviations of the eigenvalues in terms of the standard deviations of the refined parameters.

[3]In the *International Tables*, and elsewhere in the literature, use is made of the matrix $\mathbf{U}'' = \mathbf{G}\boldsymbol{\beta}/2\pi^2$. This matrix is related to $\mathbf{U}'$ by the similarity transformation $\mathbf{U}'' = \mathbf{A}^T\mathbf{U}'(\mathbf{A}^T)^{-1}$, and it has the same eigenvalues as $\mathbf{U}'$. $\mathbf{U}''$ is *not*, however, symmetric, and thus the expressions involved in computing eigenvalues are more complex than those for $\mathbf{U}'$.

The Equivalent Isotropic Temperature Factor

Crystallographic literature often refers to an *isotropic temperature factor*, which is one for which the ellipsoid of equal probability density has been constrained to be a sphere. When the ellipsoid has not been so constrained, it is often useful, for purposes of comparison, to compute an equivalent isotropic temperature factor. This is defined by $B = (B_{11} + B_{22} + B_{33})/3$. Also, it is useful, when preliminary stages of refinement have been carried out using isotropic temperature factors, to know what anisotropic temperature factor tensor corresponds to the equivalent sphere. This is given by $\beta_{ij} = BG_{ij}^{-1}/4$.

Effect of Symmetry

If an atom lies on a special position with a point-group symmetry other than $\bar{1}$, there are restrictions on the elements of the anisotropic temperature factor tensor, and the number of independent elements will be less than six. For example, suppose the atom lies on twofold rotation axis parallel to the **b** crystallographic axis. The temperature factor for the *hkl* reflection must be identical to that for the $\bar{h}k\bar{l}$ reflection or, in other words, the anisotropic temperature factor tensor must be invariant under a twofold rotation. This means that

$$\begin{pmatrix} \beta'_{11} & \beta'_{12} & \beta'_{13} \\ \beta'_{12} & \beta'_{22} & \beta'_{23} \\ \beta'_{13} & \beta'_{23} & \beta'_{33} \end{pmatrix} = \begin{pmatrix} -1 & 0 & 0 \\ 0 & 1 & 0 \\ 0 & 0 & -1 \end{pmatrix} \begin{pmatrix} \beta_{11} & \beta_{12} & \beta_{13} \\ \beta_{12} & \beta_{22} & \beta_{23} \\ \beta_{13} & \beta_{23} & \beta_{33} \end{pmatrix} \begin{pmatrix} -1 & 0 & 0 \\ 0 & 1 & 0 \\ 0 & 0 & -1 \end{pmatrix}.$$

Expanding the matrix multiplication, we obtain the relationships $\beta'_{11} = \beta_{11}$, $\beta'_{22} = \beta_{22}$, $\beta'_{33} = \beta_{33}$, and $\beta'_{13} = \beta_{13}$. But $\beta'_{12} = -\beta_{12}$, and $\beta'_{23} = -\beta_{23}$. Since zero is the only number that is equal to its own negative, it follows that $\beta_{12} = \beta_{23} = 0$. Inspection shows that β_{22} is an eigenvalue of the resulting matrix. In general, any vector coincident with a rotation axis of order two or greater, or perpendicular to a mirror plane, must be an eigenvector for a matrix constrained by symmetry. A vector perpendicular to a rotation axis, proper or improper, of order three or greater is a *degenerate eigenvector*; any orthogonal pair of vectors lying in the plane perpendicular to the rotation axis may be chosen as coordinate axes, and the matrix is invariant to *all* rotations around the higher order rotation axis. It follows from this that, if the point-group symmetry is cubic, all vectors are eigenvectors, and all properties describable by second-rank tensors are isotropic.

It is usually a rather straightforward matter to deduce, for any special position, what the restrictions on temperature factor coefficients are. A tricky case arises, however, when the special position lies on a twofold axis parallel to the **ab** plane of a trigonal or hexagonal crystal. Referring to Figure 5.2, we consider an atom lying on a twofold axis parallel to **a**. The temperature factor must be identical for reflections whose indices are *h*, *k*, *l*

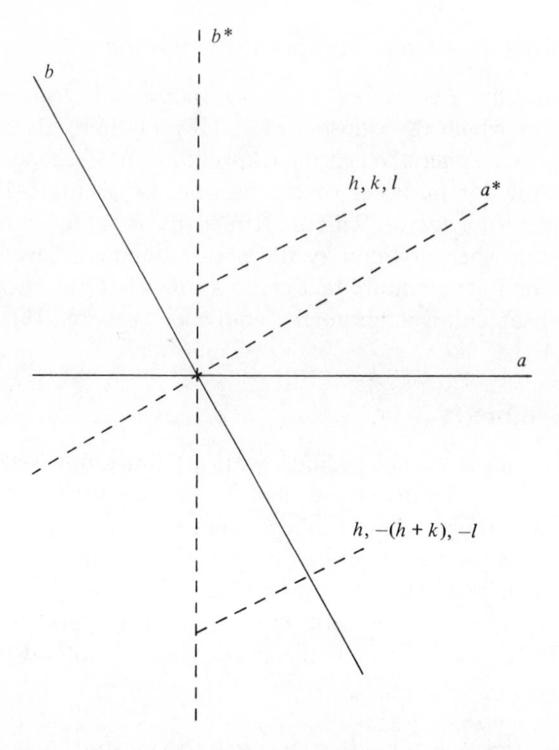

Fig. 5.2. Equivalent reflections in the hexagonal system when the **a** axis is a twofold rotation axis.

and $h, -(h+k), -l$. This gives the relationship

$$h^2\beta_{11} + k^2\beta_{22} + l^2\beta_{33} + 2hk\beta_{12} + 2hl\beta_{13} + 2kl\beta_{23}$$
$$= h^2\beta_{11} + (h^2 + 2hk + k^2)\,\beta_{22} + l^2\beta_{33} - 2(h^2 + hk)\,\beta_{12}$$
$$- 2hl\beta_{13} + 2(hl + kl)\,\beta_{23}$$

Equating the coefficients of like terms, we obtain

$$\beta_{11} = \beta_{11} + \beta_{22} - 2\beta_{12},$$

$$\beta_{22} = \beta_{22},\ \beta_{33} = \beta_{33},$$

$$2\beta_{12} = 2\beta_{22} - 2\beta_{12},$$

$$2\beta_{13} = -2\beta_{13} + 2\beta_{23},$$

$$2\beta_{23} = 2\beta_{23}.$$

From the first and fourth, we conclude that $\beta_{12} = (1/2)\beta_{22}$; from the fifth, $\beta_{23} = 2\beta_{13}$.

Tensors of Higher Ranks

We have discussed a second-rank tensor as a set of linear relationships between one vector quantity, "cause," and another vector quantity, "effect." There are physical phenomena that are themselves described by second-rank tensors. Let us consider, for example, stress and strain. A stress is a force applied to a solid body, and it can be either compressive or shearing in nature. In the cube of Figure 5.3, a force applied to the face perpendicular to axis x_1 may be either a compressive force acting parallel to x_1 or a shearing force acting parallel to x_2 or x_3. We can designate by t_{ij} a force applied to the face perpendicular to x_i in the direction parallel to x_j. Now, if a force is applied to the face perpendicular to x_1 in the direction x_2, it will generate a torque about x_3. If the cube is not to rotate, there must be an equal and opposite torque resulting from a reactive force applied to the face perpendicular to x_2 in direction x_1. Thus, for a solid body in equilibrium, $t_{ij} = t_{ji}$. Because the face of the cube is identified by a vector, the applied forces are vector quantities, and the phenomenon is independent of the orientations of any coordinate axes, the quantities t_{ij} are the elements of a second-rank symmetric tensor.

Similarily, strain can be described by a displacement of all points in a face of the cube in any of the three coordinate axis directions. The components of shear strain appear as small changes in the angles between

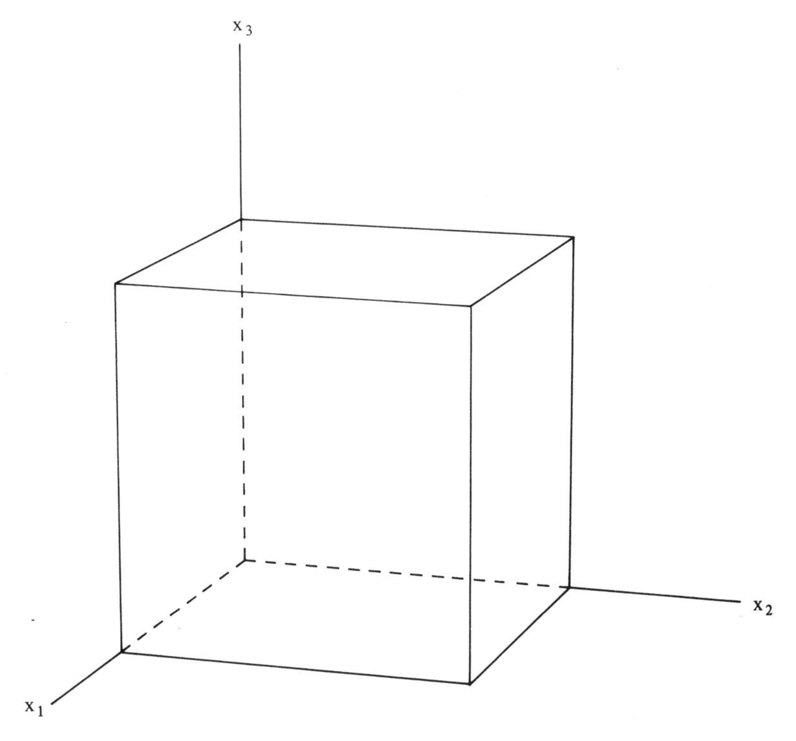

Fig. 5.3. The components of stress act on the faces of the cube.

lines that are fixed with respect to the material of the solid body and are usually defined in such a way that there is no net rotation as a result of the deformation, so that if we designate the strain tensor by **r**, then $r_{ij} = r_{ji}$, and strain is also described by a second-rank symmetric tensor.

Now let us consider the phenomenon of piezoelectricity, in which the application of a stress to a crystal produces an electric dipole moment in the crystal or, conversely, the application of an electric field to the crystal produces a strain in the crystal. If we apply a stress, **t**, to the crystal, the polarization, **p**, will be given by relationships of the form

$$p_i = \sum_{j=1}^{3} \sum_{k=1}^{3} q_{ijk} t_{jk},$$

or $\mathbf{p} = \mathbf{qt}$. The array of piezoelectric moduli, q_{ijk}, is a *tensor of the third rank*. Note that it is a three-dimensional array, so it is not a matrix, which we have defined as a two-dimensional array.

t transforms as a second-rank tensor, and **p** transforms as a vector. For the relationship between **t** and **p** to be invariant under a transformation of coordinates, **R**, **q** must transform according to

$$q'_{i'j'k'} = \sum_{i=1}^{3} \sum_{j=1}^{3} \sum_{k=1}^{3} R_{i'i} R_{j'j} R_{k'k} q_{ijk}.$$

Because the stress tensor is symmetric, $q_{ijk} = q_{ikj}$. This means that there are only 18 rather than 27 components in the general case. This fact is made use of in the conventional representation of the array of piezoelectric moduli as a 3×6 matrix. The scheme used is to designate q_{ijk} by q_{il}, where $l = j$ if $j = k$. If $j \neq k$, then take the missing index and add 3. Thus q_{i23} becomes q_{i4} (because 1 is the missing index), q_{i31} becomes q_{i5}, and q_{i12} becomes q_{i6}.

As in the case of second-rank tensors, symmetry restricts the possible values of the elements of third-rank tensors. If the point group contains the inversion, so that

$$\mathbf{R} = \begin{pmatrix} -1 & 0 & 0 \\ 0 & -1 & 0 \\ 0 & 0 & -1 \end{pmatrix},$$

it follows that $q_{ijk} = -q_{ijk}$ for all values of i, j, and k, so that piezoelectricity is impossible in any centrosymmetric point group. Appendix F gives a summary of restrictions on third-rank tensors for all other point groups.

Let us now consider elasticity, which involves relationships between stress and strain, both of which are second-rank tensors. They are related by one of the linear relationships

$$t_{ij} = \sum_{k=1}^{3} \sum_{l=1}^{3} c_{ijkl} r_{kl},$$

or

$$r_{ij} = \sum_{k=1}^{3} \sum_{l=1}^{3} s_{ijkl} t_{kl}.$$

The quantities c_{ijkl}, known as the *elastic constants* or *elastic stiffnesses*, and s_{ijkl}, known as *elastic moduli* or *elastic compliances*, are both *tensors of the fourth rank*. They transform according to the rule

$$c_{i'j'k'l'} = \sum_{i=1}^{3} \sum_{j=1}^{3} \sum_{k=1}^{3} \sum_{l=1}^{3} R_{i'i} R_{j'j} R_{k'k} R_{l'l} c_{ijkl}.$$

A fourth-rank tensor is a four-dimensional array, but, again, since stress and strain are both symmetric, $c_{ijkl} = c_{jikl} c_{ijlk} = c_{jilk}$. It is *not*, in general, possible to interchange indices of the first pair with indices of the second, so that, for example, $c_{ijkl} \neq c_{ikjl}$. It is common to reduce the four indices to two by the same scheme we used for piezoelectric moduli, so that c_{1111} becomes c_{11}, and c_{2323} becomes c_{44}, etc. The rules of symmetry place restrictions on independent elements. These also are summarized in Appendix F. Warning! It is also common to write the elastic constants tensor as a 6×6 matrix. Voigt, in his classic *Lehrbuch der Kristallphysik*, defined the elements of the compliance array as the elements of the inverse of this 6×6 matrix. The resulting array does not transform according to the transformation law, given above, for a fourth-rank tensor, and is, therefore, *not a tensor*! There have been various attempts to correct this,[4] but confusion still reigns supreme, and anyone using these quantities must be very careful to understand how they are defined.

Moments and Cumulants

We have already discussed the anisotropic temperature factor tensor, $\boldsymbol{\beta}$. The form of the structure factor expression that we have used can be written

$$F(\mathbf{h}) = \sum_{j=1}^{N} f_j \exp\left\{ 2\pi i \sum_{k=1}^{3} h_k x_k^j - \sum_{k=1}^{3} \sum_{l=1}^{3} h_k h_l \beta_{kl}^j \right\},$$

where f_j is the atomic scattering factor for atom j, whereas $\mathbf{x}^j$ and $\boldsymbol{\beta}^j$ are the position vector and the anisotropic temperature factor tensor for atom j. This expression is the sum of the Fourier transforms of Gaussian density functions with means at the various atom positions, $\mathbf{x}^j$. The fortunate circumstance of the result of Bloch's theorem, with its proof of the relationship between the probability distribution of the harmonic oscillator and the

[4]See W. A. Wooster, Crystal Physics. Cambridge University Press, Cambridge, England, 1938.

normal probability distribution of mathematical statistics, enables us to make use of a great deal of mathematical analysis that was worked out by statisticians totally independent of quantum mechanics.

Statisticians have a tendency to give their own names to concepts that are used in other branches of mathematics and the physical sciences. They call Fourier transforms, which they use a great deal, *characteristic functions*. The characteristic function, $\Phi(\mathbf{q})$, of a probability density function, $f(\mathbf{x})$, is defined by

$$\Phi(\mathbf{q}) = \langle \exp(i\mathbf{q} \cdot \mathbf{x}) \rangle = \int_{-\infty}^{+\infty} \exp(i\mathbf{q} \cdot \mathbf{x}) f(\mathbf{x})\, d\mathbf{x},$$

where the integration is performed over all space of however many dimensions. We can define *moments* of various degrees by relationships such as

$${}^{1}\mu_j = \langle x_j \rangle = \int_{-\infty}^{+\infty} x_j f(\mathbf{x})\, d\mathbf{x},$$

$${}^{2}\mu_{jk} = \langle x_j x_k \rangle = \int_{-\infty}^{+\infty} x_j x_k f(\mathbf{x})\, d\mathbf{x},$$

and so forth. Substituting a Taylor's series expansion for $\exp(i\mathbf{q} \cdot \mathbf{x})$ in the definition of $\Phi(\mathbf{q})$ we have

$$\Phi(\mathbf{q}) = \int_{-\infty}^{+\infty} \left[1 + i(\mathbf{q} \cdot \mathbf{x}) - (\mathbf{q} \cdot \mathbf{x})^2/2! - i(\mathbf{q} \cdot \mathbf{x})^3/3! + \cdots \right] f(\mathbf{x})\, d\mathbf{x}$$

$$= 1 + \sum_{i=1}^{n} \left\{ i\,{}^{1}\mu_j q_j - \sum_{k=1}^{n} \left[(1/2!)\,{}^{2}\mu_{jk} q_j q_k + \sum_{l=1}^{n} (i/3!)\,{}^{3}\mu_{jkl} q_j q_k q_l + \cdots \right]\right\}.$$

Therefore, we have

$${}^{1}\mu_j = (1/i) \left. \frac{\partial \Phi(\mathbf{q})}{\partial q_j} \right|_{\mathbf{q}=0},$$

$${}^{2}\mu_{jk} = (1/i^2) \left. \frac{\partial \Phi(\mathbf{q})}{\partial q_j \partial q_k} \right|_{\mathbf{q}=0},$$

etc. If the Fourier transform can be written as an analytic function, then all of the moments of the distribution function can be written explicitly by using these relations. The first moment, ${}^{1}\boldsymbol{\mu}$, is a vector, whereas ${}^{2}\boldsymbol{\mu}$, ${}^{3}\boldsymbol{\mu}$, and ${}^{4}\boldsymbol{\mu}$ are tensors of the second, third, and fourth ranks, respectively. Because the order of partial differentiation is immaterial, they are symmetric to the interchange of all indices and, in three-dimensional space, ${}^1\boldsymbol{\mu}$, ${}^2\boldsymbol{\mu}$, ${}^3\boldsymbol{\mu}$, and ${}^4\boldsymbol{\mu}$ have 3, 6, 10, and 15 independent elements.

For the Gaussian, or normal, distribution, the characteristic function takes the particularly simple form

$$\Phi(\mathbf{q}) = \exp\left[i\,{}^1\boldsymbol{\kappa}^T\mathbf{q} - (1/2)\mathbf{q}^{T2}\boldsymbol{\kappa}\mathbf{q} \right],$$

where the coefficients ${}^1\boldsymbol{\kappa}$ and ${}^2\boldsymbol{\kappa}$ are called *cumulants*. The expression for $\Phi(\mathbf{q})$ is identical, except for constant multipliers, to the expression for the contribution to the structure factor by a single atom. We can, therefore, establish a correspondence between the first two cumulants and the position vector and anisotropic temperature factor tensor.

In statistics, probability distributions that are almost, but not quite, can be described by treating the argument of the exponential in the characteristic function as the linear and quadratic terms of a series expansion of its logarithm. More general functions can then be described by adding to this expansion terms of cubic, quartic, or higher degree, the coefficients of which are the third, fourth, and higher cumulants. A characteristic function involving third and fourth cumulants has the form

$$\Phi(\mathbf{q}) = \exp\left\{ \sum_{j=1}^{n} \left[i\,{}^1\kappa_j q_j - \sum_{k=1}^{n} \left({}^2\kappa_{jk} q_j q_k + \sum_{l=1}^{n} \left\{ i\,{}^3k_{jkl} q_j q_k q_l - \sum_{m=1}^{n} {}^4\kappa_{jklm} q_j q_k q_l q_m \right\} \right) \right] \right\}.$$

Repeated partial-partial differentiation gives the moments of the probability density function that has this characteristic function in terms of the cumulants. The first few are

$${}^1\mu_j = {}^1\kappa_j$$

$${}^2\mu_{jk} = {}^2\kappa_{jk} + {}^1\kappa_j{}^1\kappa_k$$

$${}^3\mu_{jkl} = {}^3\kappa_{jkl} + {}^1\kappa_j{}^2\kappa_{kl} + {}^1\kappa_k{}^2\kappa_{jl} + {}^1\kappa_l{}^2\kappa_{jk} + {}^1\kappa_j{}^1\kappa_k{}^1\kappa_l.$$

Because each moment involves only cumulants of the same or lower degree, these equations can be solved for cumulants in terms of moments, giving

$${}^1\kappa_j = {}^1\mu_j$$

$${}^2\kappa_{jk} = {}^2\mu_{jk} - {}^1\mu_j{}^1\mu_k$$

$${}^3\kappa_{jkl} = {}^3\mu_{jkl} - {}^1\mu_j{}^2\mu_{kl} - {}^1\mu_k{}^2\mu_{jl} - {}^1\mu_l{}^2\mu_{jk} + 2\,{}^1\mu_j{}^1\mu_k{}^1\mu_l.$$

If we make the substitutions $\mathbf{h} = \mathbf{q}/2\pi$, $\mathbf{x} = {}^1\boldsymbol{\kappa}$, $\boldsymbol{\beta} = 2\pi^2\,{}^2\boldsymbol{\kappa}$, $\boldsymbol{\gamma} = (4\pi^3/3)^3\boldsymbol{\kappa}$, and $\boldsymbol{\delta} = (2\pi^4/3)^4\boldsymbol{\kappa}$, we can write a more general structure factor formula of the form

$$F(\mathbf{h}) = \sum_{i=1}^{N} f_j \exp\left\{ \sum_{k=1}^{3} \left[2\pi i h_k x_k^j - \sum_{l=1}^{3} \left(h_k h_l \beta_{kl}^j + \sum_{m=1}^{3} \left\{ i h_k h_l h_m \gamma_{klm}^j - \sum_{n=1}^{3} h_k h_l h_m h_n \delta_{klmn}^j \right\} \right) \right] \right\}.$$

Since all cumulants higher than the second vanish identically for a Gaussian distribution, the existence of nonzero values for third and fourth cumulants may be used as a measure of anharmonic forces, or of positional disorder, in a crystal. The third cumulant is a measure of *skewness*. A distribution with nonvanishing third cumulants, unlike the Gaussian distribution, is not an even function of all variables. The third cumulant is a third-rank tensor, so a center of symmetry forces it to vanish, and other symmetry elements force corresponding restrictions on its elements. The fourth cumulant is a measure of *kurtosis*, which is an indication of whether an actual distribution is more compact (negative kurtosis) or more spread out (positive kurtosis) than a Gaussian distribution.

Rigid-Body Motion

We have previously discussed Euler's theorem, which states that any combination of rotations of a rigid body that keep one point fixed is equivalent to a single rotation about some axis through a finite angle, and we have also discussed how a rotation can be described by an axial vector, with a magnitude given by the size of the angular displacement and a direction defined by the axis of rotation. Referring to Figure 5.4, we consider the displacement of a particle located at a position $\mathbf{r}$ with respect to the fixed origin due to a rotation through an angle λ about an axis making an angle θ with the vector, $\mathbf{r}$. We can describe this displacement by the axial vector, $\boldsymbol{\Lambda}$. The particle moves around a circle with a radius, r', given by

$$r' = |\mathbf{r}| \sin\theta = |\boldsymbol{\Lambda} \times \mathbf{r}|/\lambda.$$

The linear displacement is a chord of this circle, which is the vector sum of two vectors, one with a magnitude $r' \sin\lambda$ in a direction perpendicular to $\boldsymbol{\Lambda}$ and $\mathbf{r}$, and the other with magnitude $r'(1 - \cos\lambda)$, lying in the plane of $\boldsymbol{\Lambda}$ and $\mathbf{r}$. The linear displacement, $\mathbf{u}$, then, is given by

$$\mathbf{u} = (\sin\lambda/\lambda)(\boldsymbol{\Lambda} \times \mathbf{r}) + \left[(1 - \cos\lambda)/\lambda^2\right]\left[\boldsymbol{\Lambda} \times (\boldsymbol{\Lambda} \times \mathbf{r})\right].$$

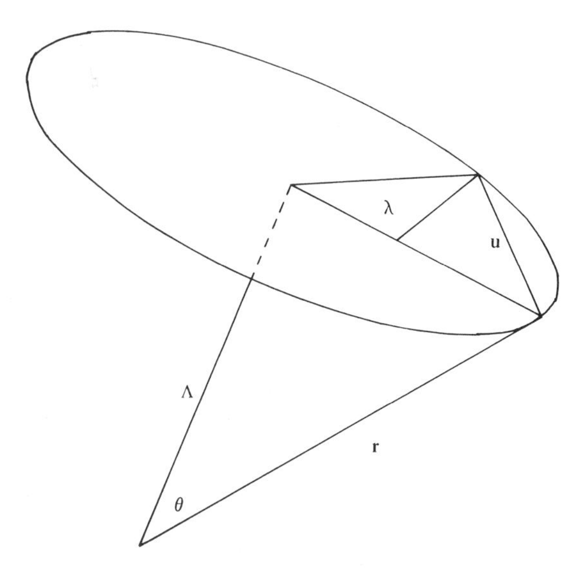

Fig. 5.4. **u** is the displacement of a particle at **r** due to the rigid-body rotation **Λ**.

This expression is exact for all finite rotations, **Λ**. For "small" displacements, however, the trigonometric functions can be approximated by their power series expansions:

$$\sin\lambda = \lambda - \lambda^3/6 + \lambda^5/120 - \cdots,$$

$$\cos = 1 - \lambda^2/2 + \lambda^4/24 - \lambda^6/720 + \cdots.$$

Because of the extremely rapid convergence of these series, due to the factorial denominators, truncation of the series after the fourth-degree terms introduces an error of $< 0.1\%$ even when λ is as large as 0.5 radian, so that, even for rather large angular displacements

$$\mathbf{u} = (1 - \lambda^2/6)(\mathbf{\Lambda} \times \mathbf{r}) + \left[(1/2) - \lambda^2/24\right]\left[\mathbf{\Lambda} \times (\mathbf{\Lambda} \times \mathbf{r})\right].$$

By expanding the vector products, and remembering that $\lambda^2 = \lambda_1^2 + \lambda_2^2 + \lambda_3^2$, the ith component of **u** can be written

$$u_i = \sum_{j=1}^{3}\left\{A(\mathbf{r})_{ij}\lambda_j + \sum_{k=1}^{3}\left[B(\mathbf{r})_{ijk}\lambda_j\lambda_k + \sum_{l=1}^{3}\left(C(\mathbf{r})_{ijkl}\lambda_j\lambda_k\lambda_l + \sum_{m=1}^{3} D(\mathbf{r})_{ijklm}\lambda_j\lambda_k\lambda_l\lambda_m\right)\right]\right\},$$

where A, B, C, and D represent coefficients of corresponding terms in the expansion. These coefficients, all of them linear functions of components of **r**, are listed in Table 5.1.

If the center of mass of the rigid body also moves, then the total displacement **v**, of a particle is **u** + **t**, where **t** is the displacement of the center of mass from its equilibrium position. When we discuss the effects of rigid-body motion on diffraction intensities, we shall be interested in the moments of **v**, which are quantities of the type $\langle v_i v_j \rangle$. These will require evaluation of quantities of the type $\langle \lambda_i \lambda_j \rangle$, $\langle t_i t_j \rangle$, and $\langle \lambda_i t_j \rangle$, as well as higher moments such as $\langle \lambda_i \lambda_j t_k \rangle$ and $\langle \lambda_i \lambda_j t_k t_l \rangle$. The second moments are used to define the *rigid body motion tensors*. The *translation tensor*, **T**, is defined by $T_{ij} = \langle t_i t_j \rangle$. The *libration tensor*, **L**, is defined by $L_{ij} = \langle \lambda_i \lambda_j \rangle$. A third tensor, **S**, defined by $S_{ij} = \langle \lambda_i t_j \rangle$, is often called the *screw correlation tensor*, although only its diagonal elements, S_{ii}, actually correspond to screwlike motions. The off-diagonal elements, S_{ij}, where $i \neq j$, correspond to rotations around axes that do not pass through the origin. Because $\langle \lambda_i t_j \rangle \neq \langle \lambda_j t_i \rangle$, the **S** tensor is not symmetric. Actually, **T**, **L**, and **S** can be viewed as partitions of a 6 × 6 matrix of the form

$$\left(\begin{array}{c|c} \mathbf{L} & \mathbf{S} \\ \hline \mathbf{S} & \mathbf{T} \end{array}\right).$$

T and **L** conform to the same symmetry restrictions that apply to the other second-rank tensors we have considered. However, because **t** is a polar vector, and Λ is an axial vector that tranforms differently under inversions and improper rotations, the symmetry restrictions on **S** are listed separately in Appendix F.

By making the assumption that the equilibrium position and the means of the probability distributions of **t** and Λ coincide, it is always possible to arrange for $\langle t_i \rangle$ and $\langle \lambda_i \rangle$ to vanish. Third moments, however, do not necessarily vanish under the same conditions. In particular, although **S** vanishes when the origin is a center of inversion, averages such as $\langle \lambda_i t_j t_k \rangle$, which are even in the components of the polar vector **t**, do not, because the axial vector Λ is invariant under inversion. A sufficient condition for $\langle \lambda_i t_j t_k \rangle$ to vanish is that the *joint distribution* of λ_i, t_j, and t_k be even in all three quantities. Although it is possible to construct a probability density function (one that is invariant under the operations of the point group 222) that is not even in all three quantities although all three marginal distributions are even, this is an improbable, special case, and, in practice, the assumption that $\langle \lambda_i t_j t_k \rangle = 0$ is equivalent to the assumption that the marginal distributions are symmetric about their means.

Bloch's theorem tells us that a distribution function for a particle is Gaussian if the potential function is harmonic. The potential function for a rigid-body rotation obviously cannot be harmonic, since it must be periodic with a period of 2π radians, but, if the energy of an oscillation is far below the barrier to rotation, we can assume that the potential function is almost

Table 5.1. Values of the Coefficients A_{ij}, B_{ijk}, C_{ijkl}, and D_{ijklm} for Computing Moments in Librating, Rigid Bodies. Arrays are Symmetric under Interchange of Indices Other Than i

$A(\mathbf{r})$ i/j	1	2	3
1	0	r_3	$-r_2$
2	$-r_3$	0	r_1
3	r_2	$-r_1$	0

$B(\mathbf{r})$ i/jk	11	12	13	22	23	33
1	0	$r_2/4$	$r_3/4$	$-r_1/2$	0	$-r_1/2$
2	$-r_2/2$	$r_1/4$	0	0	$r_3/4$	$-r_2/2$
3	$-r_3/2$	0	$r_1/4$	$-r_3/2$	$r_2/4$	0

$C(\mathbf{r})$ i/jkl	111	112	113	122	123	133	222	223	233	333
1	0	$\frac{-r_3}{18}$	$\frac{r_2}{18}$	0	0	0	$\frac{-r_3}{6}$	$\frac{r_2}{18}$	$\frac{-r_3}{18}$	$\frac{r_2}{6}$
2	$\frac{r_3}{6}$	0	$\frac{-r_1}{18}$	$\frac{r_3}{18}$	0	$\frac{r_3}{18}$	0	$\frac{-r_1}{18}$	0	$\frac{-r_1}{6}$
3	$\frac{-r_2}{6}$	$\frac{r_1}{18}$	0	$\frac{-r_2}{18}$	0	$\frac{-r_2}{18}$	$\frac{r_1}{6}$	0	$\frac{r_1}{18}$	0

$D(\mathbf{r})$ $i/jklm$	1111	1112	1113	1122	1123	1133	1222	1223	1233	1333	2222	2223	2233	2333	3333
1	0	$\frac{-r_2}{96}$	$\frac{-r_3}{96}$	$\frac{r_1}{144}$	0	$\frac{r_1}{144}$	$\frac{-r_2}{96}$	$\frac{-r_3}{288}$	$\frac{-r_2}{288}$	$\frac{-r_3}{96}$	$\frac{r_1}{24}$	0	$\frac{r_1}{72}$	0	$\frac{r_1}{24}$
2	$\frac{r_2}{24}$	$\frac{-r_1}{96}$	0	$\frac{r_2}{144}$	$\frac{-r_3}{288}$	$\frac{r_2}{72}$	$\frac{-r_1}{96}$	0	$\frac{-r_1}{288}$	0	0	$\frac{-r_3}{96}$	$\frac{r_2}{144}$	$\frac{-r_3}{96}$	$\frac{r_2}{24}$
3	$\frac{r_3}{24}$	0	$\frac{-r_1}{96}$	$\frac{r_3}{72}$	$\frac{-r_2}{288}$	$\frac{r_3}{144}$	0	$\frac{-r_1}{288}$	0	$\frac{-r_1}{96}$	$\frac{r_3}{24}$	$\frac{-r_2}{96}$	$\frac{r_3}{144}$	$\frac{-r_2}{96}$	0

harmonic near the equilibrium position and that the distribution function is therefore approximately Gaussian. Because fourth moments usually appear as correction terms in expressions that are dominated by lower order moments, it is often a justifiable assumption that the fourth moments are well-approximated by the fourth moments of a Gaussian distribution, so that $\langle x_i x_j x_k x_l \rangle = \langle x_i x_j \rangle \langle x_k x_l \rangle + \langle x_i x_k \rangle \langle x_j x_l \rangle + \langle x_i x_l \rangle \langle x_j x_k \rangle$. Thus

$$\langle \lambda_i \lambda_j \lambda_k \lambda_l \rangle = L_{ij} L_{kl} + L_{ik} L_{jl} + L_{il} L_{jk},$$

$$\langle \lambda_i \lambda_j t_k t_l \rangle = L_{ij} T_{kl} + S_{ik} S_{jl} + S_{il} S_{jk},$$

and so forth.

Chapter 6

Data Fitting

A universal problem of the experimental, physical sciences consists of asking and answering two questions. The first is, "Given a set of experimental data, y_i, and a theoretical model that establishes a connection between the data and a set of parameters, x_j, what are the values of the parameters that give the best fit to the data?" The second question is, "Having found the best fit, what can we say about the adequacy of the model in describing the data, and within what ranges do the true values of the parameters lie?" To establish a practical procedure for answering these questions, we must first find the answers to several auxiliary questions. The first, and most important, of these is, "What do we mean by the *best fit*?" We shall assume, in the following discussion, that the best fit corresponds to a minimum value of some function, $S(\mathbf{y}, \mathbf{x})$, of all data points and all parameters. In this chapter we shall begin with a discussion of the form of the function S in somewhat greater detail than usually appears in treatments of model fitting, for the purpose of highlighting some of the assumptions that are made implicitly when a particular procedure is used. We shall then discuss various approaches to the numerical analysis problem of finding the minimum of this function. In subsequent chapters we shall discuss the connected problems of assessing the precision of the results and constructing and comparing models that obey the laws of physics and chemistry.

Fitting Functions

Let us consider a set of N observations, y_i, that have been measured experimentally, each subject to some random error due to the finite precision of the measurement process. We consider that each observation is randomly selected from some population that can be described by a statistical distribution function with a mean and a variance. We then ask the question, "What is the *likelihood* that, by chance, we would have observed the particular set of values we did observe?" We may assume that the values of model parameters that maximize this likelihood will be a good *estimate* of the true values of these parameters if the model corresponds to a good description of physical reality.

In explicit terms, we assume that $y_i = M_i(\mathbf{x}) + e_i$, where $M(\mathbf{x})$ represents a model function and the e_i are random errors distributed according to some density function, $f_i(e)$. In the case of most physical measurements the value of one observation is not influenced by other observations of the same quantity, or of different quantities, so that the raw data may be assumed to be uncorrelated, and their joint distribution is therefore the product of their individual marginal distributions. The likelihood function, then, is given by

$$L = \prod_{i=1}^{N} f_i[y_i - M_i(\mathbf{x})].$$

Because f_i is a probability density function, it must be everywhere greater than or equal to zero, and thus have a real logarithm. The logarithm is a monotonically increasing function of its argument, so the maximum value of L corresponds also to the maximum value of $\ln(L)$. Therefore we have

$$\ln(L) = \sum_{i=1}^{N} \ln\{f_i[y_i - M_i(\mathbf{x})]\}.$$

Gauss considered the case where the error distribution is Gaussian: that is

$$f_i(R) = (2\pi)^{-1/2}\sigma_i^{-1}\exp\left[-(1/2)(R_i/\sigma_i)^2\right],$$

where $R_i = [y_i - M_i(\mathbf{x})]$, and σ_i^2 is the variance of the ith observation. In this case

$$\ln(L) = -(1/2)\sum_{i=1}^{N}(R_i/\sigma_i)^2 - \sum_{i=1}^{N}\ln(\sigma_i) - (N/2)\ln(2\pi).$$

The second and third terms are independent of $\mathbf{x}$, so $\ln(L)$ has its maximum value when

$$S = \sum_{i=1}^{N}(R_i/\sigma_i)^2 \quad \text{is a minimum.}$$

Therefore, *if the error distributions are Gaussian, and observations are weighted by the reciprocals of their variances, the method of least squares gives the maximum likelihood estimate of the parameters,* x_j.

In practice the actual shape of the error distribution function and its variance are usually unknown, so we must investigate the consequences if the conditions stated above are not met. In general, the method of least squares does not lead to a maximum likelihood estimate. To consider an extreme case, if f_i were a Cauchy distribution,

$$f(x) = \left\{(\pi s)\left[1 + (x/s)^2\right]\right\}^{-1},$$

then $-\ln(L)$ would be the slowly rising function illustrated in Figure 6.1. (Here s is a measure of scale which should not be confused with σ. A Cauchy distribution has no variance, because

$$\int_{-\infty}^{+\infty} x^2\left[1 + (x/s)^2\right]^{-1} dx$$

is infinite. Fortunately, distributions as long-tailed as a Cauchy distribution are rarely, if ever, encountered in nature.) In spite of the fact that least squares is not optimal, there is justification for using it in cases where the conditions are only approximately met. In particular, the *Gauss-Markov theorem* states that, if the errors are random and uncorrelated, the method of least squares gives the *best linear unbiased estimate* of the parameters, meaning that of all functions for which each parameter is a linear function of the data points, least squares is the one for which the variances of the parameters are smallest.

Nevertheless, if the tails of the experimental error distribution contain a substantially larger proportion of the total area than the tails of a Gaussian distribution, the "best linear" estimate may not be very good, and there will usually be a procedure in which the parameters are nonlinear functions of the data that gives lower variances for the parameter estimates than does least squares. With this in mind, let us consider two general properties of fitting algorithms, *robustness* and *resistance*.

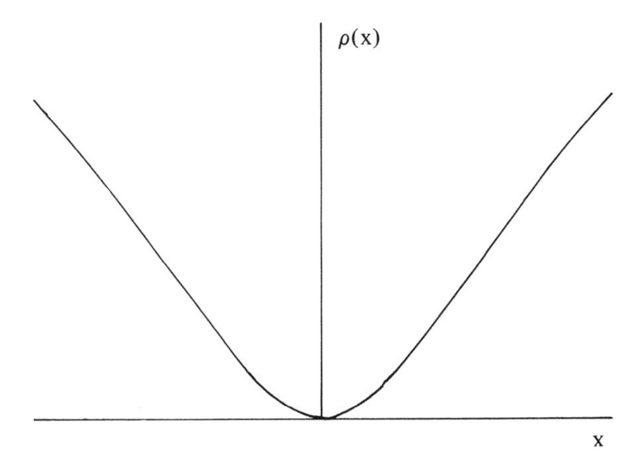

Fig. 6.1. The fitting function that leads to the maximum likelihood estimate of x if the errors in the observations have a Cauchy distribution. $\rho(x) = \ln(1 + x^2)$.

A procedure is said to be *robust* if it gives parameter estimates with variances close to the minimum variance for a wide range of error distributions. Least squares is very sensitive to the effects of large residuals, so the results are distorted if large differences between the observed data and the model predictions are present with frequencies substantially greater than those in a Gaussian distribution. Least squares is therefore not robust. A procedure is *resistant* if it is insensitive to the presence or absence of any small subset of the data. Although resistance in general applies to *any* small subset of the data, in practice it applies particularly to small numbers of data points that are wildly discrepant relative to the body of the data—so-called "outliers." There are several reasons why data may be discrepant, a gross error of measurement being only the most obvious. Another is the fact that certain data points may be particularly sensitive to some unmodeled (or inadequately modeled) parameter, or, from another point of view, particularly sensitive to some systematic error that has not been accounted for in the experiment.

In recent years a great deal of work has been done on determining what properties a robust and resistant procedure should have. Obviously, if the error distribution is Gaussian, or very close to Gaussian, the procedure should give results very close to those given by least squares. This suggests that the procedure should be much like least squares for small values of the residuals. Because the weakness of least squares lies in its overemphasis of large values of the residuals, these should be deemphasized, or perhaps even ignored. For intermediate values of the residuals the procedure should connect the treatments of the small residuals and the extremely large residuals in a smooth fashion.

To define a robust/resistant procedure we shall state the fitting problem in the somewhat more general terms: "Find the minimum value of a function of a vector, $\mathbf{x}$, of adjustable, model parameters, defined by

$$f(\mathbf{x}) = \sum_{i=1}^{N} \rho(R_i/\sigma_i),$$

where $R_i = [y_i - M_i(\mathbf{x})]$ is the difference between the observed and calculated values of the ith data point, and σ_i is its estimated (or perhaps we should call it 'guesstimated') standard deviation." A necessary, but not in general sufficient, condition for $f(\mathbf{x})$ to be a minimum is for the gradient to vanish, giving a system of equations such as

$$\frac{\partial f(\mathbf{x})}{\partial x_j} = \sum_{i=1}^{N} \phi[R_i(\mathbf{x})/\sigma_i] R_i(\mathbf{x}) \frac{\partial M_i(\mathbf{x})}{\partial x_j} = 0.$$

Here $\phi(y) = (1/y)(d/dy)\rho(y)$. If $\rho(y) = y^2/2$, corresponding to least squares, $\phi(y) = 1/\sigma_i^2$, and the response is a linear function of each R_i. If the experimental errors are distributed according to a probability density function, $\Psi(y)$, and $\rho(y) = \ln\Psi(y)$, the minimum corresponds to the maximum likelihood estimate of $\mathbf{x}$. Therefore, if we knew the form of the function $\Psi(y)$ for a particular experiment, we could tailor-make a function $\phi(y)$ that would maximize the likelihood of obtaining the particular set of values we did get in that experiment.

Although we rarely, if ever, really know the form of $\Psi(y)$, it is still a legitimate procedure to propose a function, $\phi(y)$, having the properties we have outlined for a robust/resistant procedure, and to examine its behavior under more or less realistic conditions. $\phi(y)$ need not correspond to the maximum likelihood estimator for any $\Psi(y)$ to be a good estimator for a range of density functions. One form of $\phi(y)$, proposed by J. W. Tukey, is

$$\begin{aligned} \phi(y) &= \left[1-(y/a)^2\right]^2 && \text{for } |y| \leqslant a, \\ \phi(y) &= 0 && \text{otherwise,} \end{aligned}$$

corresponding to the function, $\rho(y)$,

$$\begin{aligned} \rho(y) &= (y^2/2)\left[1-(y/a)^2+(1/3)(y/a)^4\right] && \text{for } |y| \leqslant a, \\ \rho(y) &= a^2/6 && \text{otherwise.} \end{aligned}$$

Here a is a measure of scale chosen so as to exclude, at most, a small number of extreme data points. Another form for $\phi(y)$, proposed by D. F. Andrews, is

$$\begin{aligned} \phi(y) &= \sin(y/a)/(y/a) && \text{for } |y| \leqslant \pi a, \\ \phi(y) &= 0 && \text{otherwise,} \end{aligned}$$

corresponding to

$$\begin{aligned} \rho(y) &= 1-\cos(y/a) && \text{for } |y| \leqslant \pi a, \\ \rho(y) &= 2 && \text{otherwise.} \end{aligned}$$

These functions are illustrated in Figure 6.2. Both $\rho(y)$ functions lie close to a parabola for small $|y|$, and are constant for large $|y|$, so that the function $f(\mathbf{x})$ behaves like least squares if the residual is small, but large residuals are ignored.

The function $\phi(y)$ appears in the expressions for the components of the gradient of $f(\mathbf{x})$ in the position where the weights appear in least squares. (For least squares, $\phi(R_i/\sigma_i) = w_i = 1/\sigma_i^2$.) For this reason the robust/resistant procedure is sometimes described in terms of modifying the weights in each cycle of an iterative procedure for finding the minimum and, therefore, the procedure is sometimes called *iteratively weighted least squares*. It should be noted, however, that the function that is minimized is really a

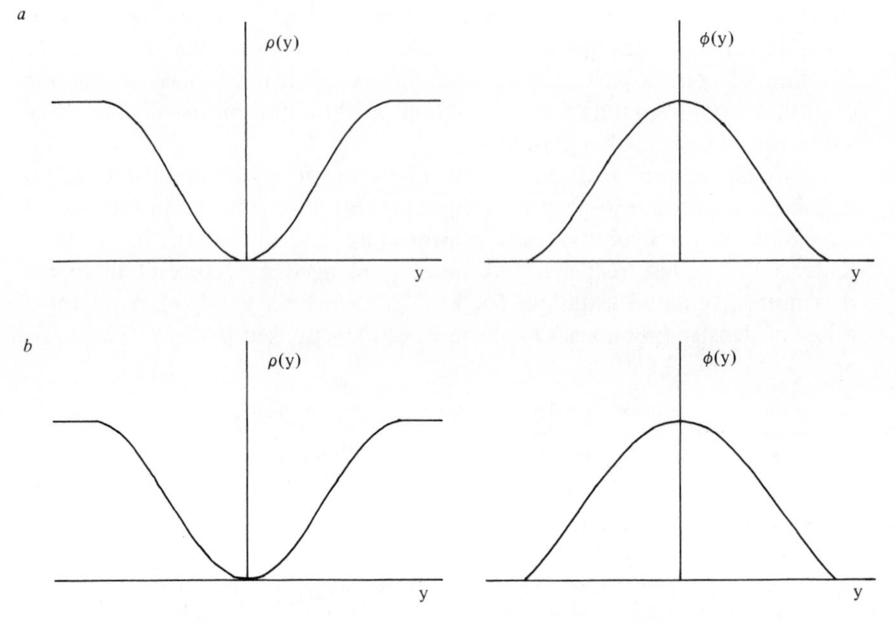

Fig. 6.2. Two robust/resistant fitting functions. (a) Tukey's function: $\rho(y) = (y^2/2) \cdot [1 - (y/a)^2 + (1/3)(y/a)^4]$; $\phi(y) = [1 - (y/a)^2]^2$. (b) Andrews' function: $\rho(y) = 1 - \cos(y/a)$; $\phi(y) = \sin(y/a)/(y/a)$.

more complex function of the residuals than a sum of squares, so this designation is no more than suggestive.

Finding the Minimum

As we have seen, a necessary condition for finding the minimum of the function

$$f(\mathbf{x}) = \sum_{i=1}^{N} \rho(R_i/\sigma_i)$$

is to find a solution of a system of equations of the type

$$\sum_{i=1}^{N} \phi\left[R_i(\mathbf{x})/\sigma_i\right] R_i(\mathbf{x}) \frac{\partial R_i(\mathbf{x})}{\partial x_j} = 0.$$

In crystallographic structure refinement, and in many other common problems, $R_i(\mathbf{x})$ is a transcendental function of the adjustable parameters, x_j, and there is no direct algorithm for finding a solution of the system of equations. One is forced, therefore, to use an indirect algorithm, and most of these are variations of Newton's method for finding the roots of

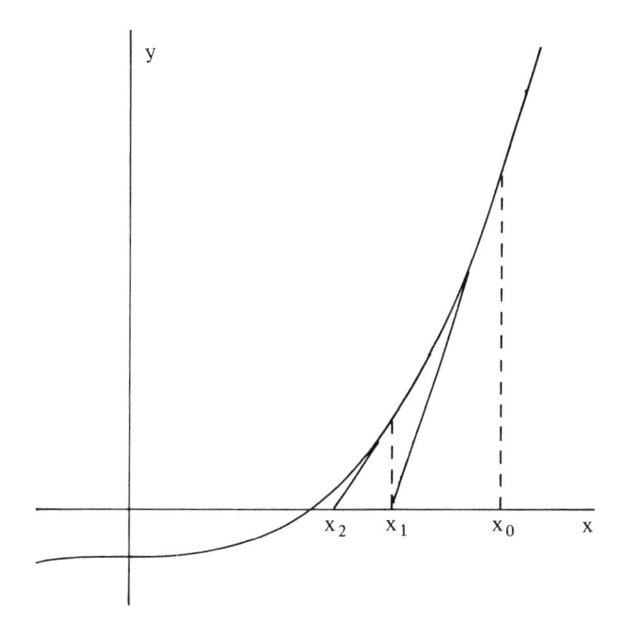

Fig. 6.3. Newton's method for finding the roots of nonlinear equations.

nonlinear equations. In Newton's method (see Fig. 6.3) we guess at an approximate solution, x_0, and find the tangent to the curve $y = f(x)$ at the point $[x_0, f(x_0)]$. The tangent line is the linear function $y = f(x_0) + (x - x_0) \cdot f'(x_0)$, where $f'(x)$ denotes $(d/dx)f(x)$. Our second guess is the root of this linear function:

$$x_1 = x_0 - \left[f(x_0)/f'(x_0) \right].$$

We continue to make new guesses,

$$x_{m+1} = x_m - \left[f(x_m)/f'(x_m) \right],$$

until $f(x_m)$ is deemed to be adequately close to zero.

If the function, $f(x)$, is actually the one depicted in Figure 6.3, this procedure will converge rapidly to a stable solution. But consider the function shown in Figure 6.4, together with the rather unfortunate choice of a starting guess. Although the function has a root (at $x = -2.1669$), the process we have described will oscillate indefinitely between $x_m = 0$ for n even and $x_m = \sqrt{3/2}$ for n odd. It must be recognized, therefore, that most algorithms are subject to pathological conditions in which they converge slowly, if at all.

In many-parameter problems Newton's method consists of making similar linear approximations to each of the nonlinear functions that describe

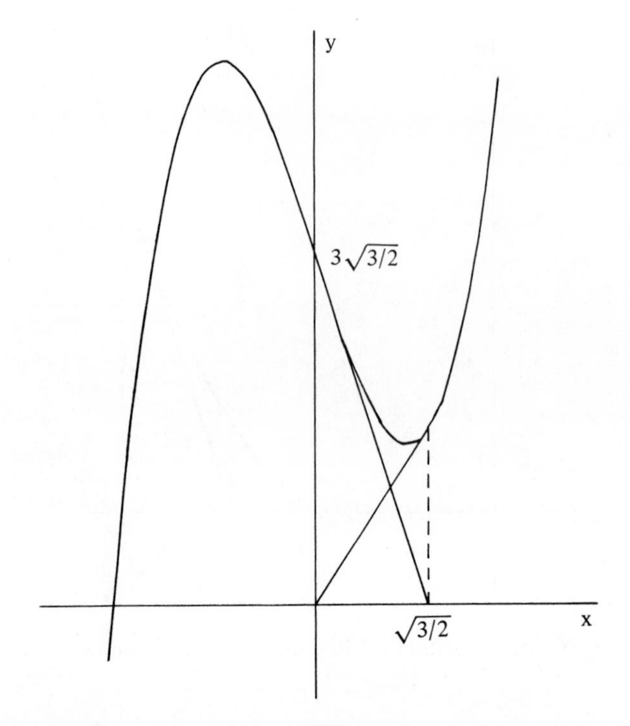

Fig. 6.4. The function $y = x^3 - 3x + 3\sqrt{3/2}$. If the initial guess is $x = 0$, Newton's method will cycle indefinitely, with $x_n = 0$ for even values of n and $x_n = \sqrt{3/2}$ for odd values of n.

the gradient, which produces a system of linear equations of the type

$$\sum_{i=1}^{N}\left\{\phi[R_i(\mathbf{x}_0)/\sigma_i]R_i(\mathbf{x}_0)\frac{\partial R_i}{\partial x_k}\right.$$

$$+\sum_{i=1}^{P}\left[\omega(R_i\{\mathbf{x}_0\}/\sigma_i)\frac{\partial R_i}{\partial X_j}\frac{\partial R_i}{\partial X_k}\right.$$

$$\left.\left.+\phi(R_i\{\mathbf{x}_0\}/\sigma_i)R_i(\mathbf{x}_0)\frac{\partial^2 R_i}{\partial X_j \partial X_k}\right]\Delta x_k\right\} = 0,$$

where $\omega(y) = (d/dy)[y\phi(y)] = (d^2\rho(y)/dy^2)$. Because this system of equations is linear, it can be written more conveniently in matrix form. Let us designate by $\mathbf{Y}$ the N element column vector whose elements are $Y_i = -\phi(R_i/\sigma_i)R_i$, and let $\mathbf{A}$ be the $N \times P$ matrix, the *design matrix*, whose elements are $A_{ij} = (\partial R_i / x_j)$. The matrix $\mathbf{H}$, whose elements are

$$H_{jk} = \frac{\partial^2 f(\mathbf{x})}{\partial x_j \partial x_k} = \sum_{i=1}^{N} \left\{ \omega\left[R_i(\mathbf{x}_0)/\sigma_i \right] \frac{\partial R_i}{\partial x_j} \frac{\partial R_i}{\partial x_k} \right.$$

$$\left. + \phi\left[R_i(\mathbf{x}_0)/\sigma_i \right] R_i(\mathbf{x}_0) \frac{\partial^2 R_i}{\partial x_j \partial x_k} \right\},$$

is called the *Hessian matrix* or the *normal equations matrix* of the function $f(\mathbf{x})$. If we designate by $\mathbf{z}$ the vector $(\mathbf{x}_1 - \mathbf{x}_0)$, Newton's method consists of solving the matrix equation $\mathbf{H}\mathbf{z} = \mathbf{A}^T\mathbf{Y}$, for which the solution is $\mathbf{z} = \mathbf{H}^{-1}\mathbf{A}^T\mathbf{Y}$.

As in the case of Newton's method with one variable, if this method is applied repeatedly, the magnitude, $(\mathbf{z}^T\mathbf{z})^{1/2}$, of the vector $\mathbf{z}$ will usually approach zero fairly rapidly. However, if the vector $\mathbf{x}_0$ is too far away from the vector, $\mathbf{x}_m$, that corresponds to the minimum of $f(\mathbf{x})$, the procedure may fail to converge. Further, because $f(\mathbf{x})$ is often a highly complex function, there is no guarantee that the point at which the gradient vanishes that is determined by this procedure is the true *global minimum*. It may be a higher *false minimum*, or even, though this is unlikely, a *saddle point*. There is no purely mathematical way to avoid a false minimum, so the problem is, strictly speaking, beyond the scope of this book but, as we shall discuss later, there are certain things that can be done to reduce the susceptibility of a given problem to falling into this trap.

In actual practice with the use of Newton's method it is usual to omit the term involving the second partial derivatives from the expression for an element of the Hessian matrix. This practice is reasonably well-justified for two reasons. The main reason is that, in the region close to the solution, the model function, $M_i(\mathbf{x})$, can be approximated by a power series expansion in the parameters in which only linear terms are retained:

$$M_i(\mathbf{x}) = M_i(\mathbf{x}_0) + \sum_{i=1}^{P} \frac{\partial M_i}{\partial x_j}(x_j - x_{0j}).$$

In this linearized form of the model function second derivatives vanish, so that, to the extent that this expansion is an adequate approximation, they may be ignored. Further, the second derivative term is a function of R_i. Even if the algorithm is least squares, with $\phi(R_i/\sigma_i) = 1$, only a small fraction of the values of R_i will both have appreciable magnitude and be paired with large values of the second derivative. If some robust/resistant function is used for $\phi(R_i/\sigma_i)$, ϕ will be small when R is large, and vice versa. Overall, then, terms of the form $\phi(R_i/\sigma_i)R_i(\partial^2 R_i/\partial x_j \partial x_k)$ will usually have a small effect on H_{jk}, and may therefore be safely ignored.

Another useful approximation, if robust/resistant techniques are being used, is to replace the individual values of $\omega(R_i/\sigma_i)$ for each data point by the average value of this function over the entire data set. If we designate this quantity by g,

$$g = (1/N)\sum_{i=1}^{N}\omega(R_i/\sigma_i).$$

Including these approximations, $\mathbf{H}$ reduces to the simple form $\mathbf{H} = g(\mathbf{A}^T\mathbf{A})$. If $\mathbf{H}$ has this form, the individual elements of $\mathbf{H}$ take the form

$$H_{jk} = g\sum_{i=1}^{N}\frac{\partial R_i}{\partial x_j}\frac{\partial R_i}{\partial x_k}.$$

If we designate by $\mathbf{B}_j$ the vector whose elements are $\partial R_i/\partial x_j$, we can then express the elements of $\mathbf{H}$ in the matrix form $H_{jk} = g\mathbf{B}_j^T\mathbf{B}_k$. A relationship known as Schwarz's inequality[1] states that $|\mathbf{B}_j^T\mathbf{B}_k| \leqslant [(\mathbf{B}_j^T\mathbf{B}_j)(\mathbf{B}_k^T\mathbf{B}_k)]^{1/2}$, with the equality applying only if $\mathbf{B}_j$ and $\mathbf{B}_k$ are *linearly dependent*, i.e., if the elements of one are a constant multiple of the elements of the other. Because of this the submatrix of $\mathbf{H}$

$$\begin{pmatrix} H_{jj} & H_{jk} \\ H_{jk} & H_{kk} \end{pmatrix}$$

is positive definite if $\mathbf{B}_j$ and $\mathbf{B}_k$ are linearly independent. In general, any matrix of the form $\mathbf{A}^T\mathbf{A}$ is positive definite if $\mathbf{A}$ has at least as many rows as columns, and if the columns of $\mathbf{A}$ are linearly independent.

We have a set of observed data, y_i, collected with various values of a controlled variable, x_i.

i	x_i	y_i
1	0.062	0.355
2	0.545	0.894
3	1.464	0.869
4	2.730	1.141
5	4.218	1.598
6	5.782	1.764
7	7.270	1.686
8	8.536	2.091
9	9.455	2.492
10	9.938	2.440

[1] A proof of Schwarz's inequality appears in G. W. Stewart, Introduction to Matrix Computations. Academic Press, New York, London, 1973, p. 165.

We believe that the data measure an effect that may be described by a polynomial law. We shall therefore try to fit them with the function

$$y = b_0 + b_1 x + b_2 x^2.$$

The design matrix is

$$\mathbf{A} = \begin{pmatrix} 1.0 & 0.062 & 0.0038 \\ 1.0 & 0.545 & 0.2970 \\ 1.0 & 1.464 & 2.1433 \\ 1.0 & 2.370 & 7.4529 \\ 1.0 & 4.218 & 17.4529 \\ 1.0 & 5.782 & 33.4315 \\ 1.0 & 7.270 & 52.8529 \\ 1.0 & 8.536 & 72.8633 \\ 1.0 & 9.455 & 89.3970 \\ 1.0 & 9.938 & 98.7638 \end{pmatrix},$$

and the normal equations are $\mathbf{A}^T\mathbf{A}\mathbf{b} = \mathbf{A}^T\mathbf{y}$, *or*

$$\begin{pmatrix} 10.0 & 50.0 & 375.0 \\ 50.0 & 375.0 & 3125.0 \\ 375.0 & 3125.0 & 27344.0 \end{pmatrix} \begin{pmatrix} b_0 \\ b_1 \\ b_2 \end{pmatrix} = \begin{pmatrix} 15.33 \\ 99.75 \\ 803.27 \end{pmatrix}.$$

We saw, in Chapter 1, *that the solution to these questions is* $b_0 = 0.5670$, $b_1 = 0.2185$, $b_2 = -0.00337$.

It is considered to be good practice for the number of observations, N, to be much larger than the number of parameters, P. Let us designate N/P by k. When we do the computations to find the minimum value of $f(\mathbf{x})$, we need to store only one triangle of $\mathbf{H}$, since $\mathbf{H}$ is symmetric. $\mathbf{H}$ will then have $P(P+1)/2$ independent elements, each of which requires kP multiplications and additions to compute it, giving a total of $kP^2(P+1)$ operations. If P is large, computer time increases in proportion to P^3, and, if many iterations are required to achieve convergence, the costs of a large problem such as the refinement of a protein structure can get rapidly out of control. For this reasons it is interesting to consider procedures for which it is not necessary to construct and invert the entire Hessian matrix in each iteration.

There are several procedures that depend on use of an approximation to the Hessian matrix, and compute the actual value of $f(\mathbf{x})$ additional times to avoid repeated computation of the entire matrix. These procedures usually use an initial computation of the gradient and the approximation to the Hessian matrix to determine a direction in parameter space that leads to a lower value of $f(\mathbf{x})$, and make a *line search* along that direction to find the minimum. If the Hessian matrix is approximated by a diagonal matrix of the form $H_{jj} = \mathbf{B}_j^T\mathbf{B}_j$, $H_{ij} = 0$ for $j \neq k$, the line follows the negative of the gradient at the point $\mathbf{x} = \mathbf{x}_0$, and the procedure is known as the *method of steepest descents*. In this procedure one first finds the vector, $-\mathbf{g}_0$, whose elements are $-g_{0j} = \partial f(\mathbf{x})/\partial x_j$. Then one may simply evaluate $f(\mathbf{x} - \alpha\mathbf{g}_0)$, where α is some scalar, for various values of α until a minimum is found. Or one may fit the three lowest points to a parabolic function of α, using the minimum of that function as the next guess until no lower point can be found. Another alternative is to evaluate the gradient at the point $\mathbf{x} - \alpha\mathbf{g}$, and determine the derivative of $f(\mathbf{x})$ with respect to α. If we designate the gradient at the point $\mathbf{x}_0 - \alpha\mathbf{g}_0$ by $\mathbf{g}_1$, this derivative is given by $df(x)/d\alpha = \mathbf{g}_1^T\mathbf{g}_0$, and the minimum of the function along $-\mathbf{g}_0$ is the point at which this quantity vanishes. This may be found by a process of linear interpolation and extrapolation. When the value of α is found that gives the minimum value of $f(\mathbf{x})$, the gradient at this point is used for a new line search, and so forth until no further reduction is possible.

In the method of steepest descents each search line $\mathbf{g}_{i+1}$, is *orthogonal* to the previous line, $\mathbf{g}_i$, i.e., $\mathbf{g}_{i+1}^T\,\mathbf{g}_i = 0$. There is a variation of this in which the search direction is required to be orthogonal not to the previous search line but to the vector representing the *difference* between the gradient at $\mathbf{x}_i$, $\mathbf{g}_i$, and the gradient at the previous point, $\mathbf{g}_{i-1}$. Designating the search direction for the ith step by $\mathbf{s}_i$, we have

$$(\mathbf{g}_i - \mathbf{g}_{i-1})^T\mathbf{s}_i = 0.$$

$\mathbf{s}_i$ can be represented as the sum of $-\mathbf{g}_i$ and a component parallel to $\mathbf{s}_{i-1}$, giving us $\mathbf{x}_i = \mathbf{g}_i + \gamma_i\mathbf{s}_{i-1}$, and the orthogonality condition gives $\gamma_i = (\mathbf{g}_i - \mathbf{g}_{i-1})^T\mathbf{g}_i/(\mathbf{g}_i - \mathbf{g}_{i-1})^T\mathbf{s}_{i-1}$. The importance of this method arises from the fact that, if the function $f(\mathbf{x})$ is *quadratic*, that is if $f(\mathbf{x}) = \mathbf{x}^T\mathbf{H}\mathbf{x}$, where $\mathbf{H}$ is a constant positive definite matrix, it can be shown that $\mathbf{g}_i^T\mathbf{g}_j = 0$ for *all* $j < i$. This method is known, therefore, as the *method of conjugate gradients*. Because, in P dimensions, there can be no more than P mutually orthogonal vectors, the vector $\mathbf{s}_{p+1}$ can only be the null vector, and the method is guaranteed to find the minimum of a quadratic function in no more than P line searches. Figure 6.5 shows what happens in two dimensions. The method of steepest descents will get closer to the minimum in ever smaller steps, but the method of conjugate gradients finds the minimum on the second try.

The obvious weakness of the method of conjugate gradients is that the convergence properties that apply if the function, $f(\mathbf{x})$, is quadratic *do not* apply if, as is usually the case, the function *is not* quadratic. The method, nevertheless, has the virtue that the Hessian matrix need never be calculated, which can be a big advantage if the number of parameters, P, is large. A further point is that any function that possesses a minimum can be expanded in a Taylor's series in $(\mathbf{x} - \mathbf{x}_m)$, where $\mathbf{x}_m$ denotes the minimum, and terms of higher degree than quadratic can be neglected for $\mathbf{x}$ close

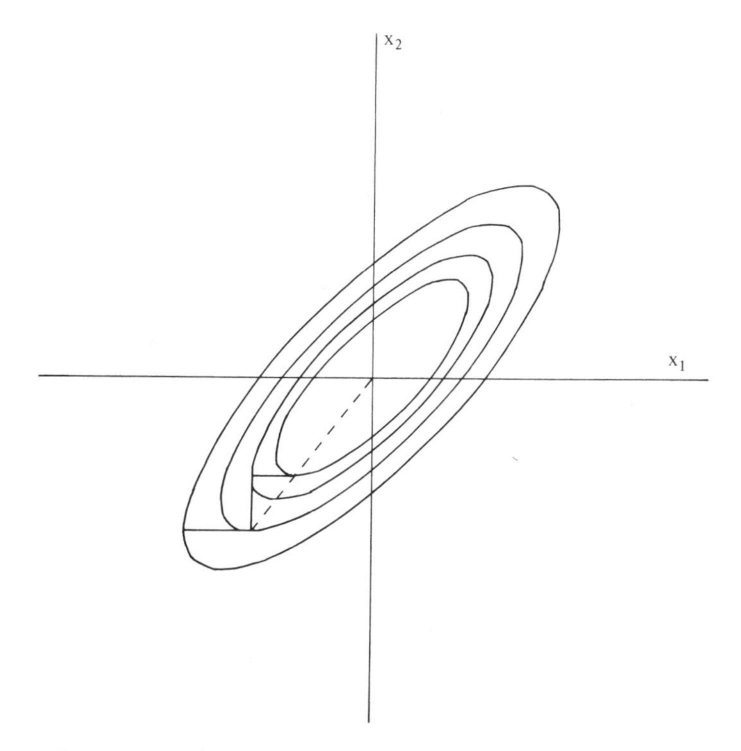

Fig. 6.5. The contours of a quadratic function in two dimensions. The method of steepest descents (*solid lines*) cycles ever closer to the minimum, but never reaches it. The method of conjugate gradients, however, follows the *dashed line* on the second step and goes directly to the minimum.

enough to $\mathbf{x}_m$. The recommended procedure is to start over with $\mathbf{s}_i = -\mathbf{g}_i$ for every $i = P + 1$ steps. It will always converge, and it will converge rapidly when $\mathbf{x} - \mathbf{x}_m$ is small enough for the quadratic approximation to be a good one.

Another alternative to Newton's method for general functions involves, again, starting with an approximation to the Hessian matrix or, rather, the inverse of the Hessian matrix. The approximation can be the full matrix for the linearized problem, in which case it begins in the same way as the simplified Newton procedure described earlier, or it can be the diagonal approximation, in which case the procedure begins with a steepest descents step. In either case a line search is made along the line in parameter space that is predicted, by the approximate Hessian matrix, to lead to the global minimum. The function will have a minimum along this line, but it (1) will not be the distance along the line that was predicted and (2) will not have a vanishing gradient at that point. The actual point at which the minimum is found and the actual gradient at that point are used to calculate corrections to the inverse Hessian matrix to give one that would have predicted the values actually found. A new prediction is then made, using the corrected matrix, and the procedure is repeated until a minimum is found. Because this procedure resembles Newton's method except that the Hessian matrix is calculated numerically rather than analytically, its general class is referred to by the name *quasi-Newton methods*.

Many quasi-Newton algorithms have been proposed that differ from one another in the way $\mathbf{H}^{-1}$ is changed at the end of each step. A particular, widely used example of this class is known, after its developers, as the Davidon-Fletcher-Powell, or DFP, algorithm.[2] If we designate $\mathbf{H}^{-1}$ by $\mathbf{G}$, $(\mathbf{x}_{i+1} - \mathbf{x}_i)$ by $\mathbf{s}_i$, and $(\mathbf{g}_{i+1} - \mathbf{g}_i)$ by $\mathbf{y}_i$, then the DFP algorithm uses the relationship

$$\mathbf{G}_{i+1} = \mathbf{G}_i - \frac{\mathbf{G}_i\mathbf{y}_i\mathbf{y}_i^T\mathbf{G}_i}{\mathbf{y}_i^T\mathbf{G}_i\mathbf{y}_i} + \frac{\mathbf{s}_i\mathbf{s}_i^T}{\mathbf{s}_i^T\mathbf{y}_i}$$

to update the matrix. Note that, if $f(\mathbf{x})$ is quadratic, and $\mathbf{H}$ is the true Hessian matrix, then $\mathbf{s}_1 = -\mathbf{G}_1\mathbf{g}_1$, and $\mathbf{g}_2 = 0$, so that $\mathbf{y}_1 = \mathbf{g}_1$, from which it follows that $\mathbf{G}_2 = \mathbf{G}_1$. Thus, if $\mathbf{H}$ is close to being the correct Hessian matrix, and $(\mathbf{x}_i - \mathbf{x}_m)$ is small enough for the quadratic approximation to be reasonable, this algorithm converges rapidly, with small corrections, to the minimum.

All of these alternatives to Newton's method for finding the minimum represent tradeoffs in which greater stability, along with the avoidance of the repeated construction of a large matrix, are bought at the expense of a larger number of times that the value of $f(\mathbf{x})$ and its derivatives must be computed. In general they are most advantageous if the number of parameters, P, is large. They can be advantageous, however, even when P is only moderately large, if it is difficult to make close guesses for the initial values of some important parameters. That condition can cause the straight use of Newton's method to be unstable.

False Minima

All of the numerical procedures we have discussed are designed to find points in a multidimensional parameter space at which the gradient of a function, $f(\mathbf{x})$, vanishes. In the case of a quadratic function of the form $f(\mathbf{x}) = \mathbf{s}^T\mathbf{H}\mathbf{s}$, where $\mathbf{s}$ represents $(\mathbf{x} - \mathbf{x}_m)$, and $\mathbf{H}$ is a positive definite matrix, there will be only one such point, and it must be a relative minimum. In crystallographic refinement, and in many other practical, data-fitting problems, the model function is usually more complex than quadratic, and it is frequently transcendental. It may, therefore, have many

[2] A thorough discussion of this and many other methods of finding the minimum of a general function appears in W. Murray (ed.) Numerical Methods for Unconstrained Optimization. Academic Press, London, New York, 1972.

stationary points, points at which the gradient vanishes and, although these may include relative maxima and saddle points, some are likely to be relative minima with values of $f(\mathbf{x})$ greater than the value at the true *global minimum*. There is no mathematical way to assure that the minimum found by the computer is not one of these *false minima*. There are, however, some general principles that may be followed to reduce the chances of being caught in such a trap, and a few "tricks of the trade" that may be used to escape from the trap if it is encountered.

It is reasonable to conjecture that the global minimum corresponds to a situation in which the value predicted by the model for a data point is close to the observed value for all, or almost all, of the data points. For each data point there will be other values of the parameters that also predict a value close to the observed value, but it is likely that these other parameter values will be different for different data points. Further, there may be values of the parameters for which the differences between the predicted values and the observed values for some data points are at relative minima, although they may not be small. False minima result when either the secondary agreement points occur in the same neighborhood for a substantial subset of the data, or a few data points have relative minima in their differences at large values of those differences, so that a small number of points makes an inordinately large contribution to the function. In either case the false minimum occurs because of the existence of a relative minimum for some *subset* of the data.

This conjecture suggests several ways to improve one's chances of avoiding false minima. One is to collect much more independent data than are necessary to give a unique solution to the problem. The hope is to make the number of distinct relationships contributing to the sum, $f(\mathbf{x})$, large enough that only a small fraction of them will have subsidiary minima in any region of parameter space other than the vicinity of the global minimum. A second strategy is to use a resistant function. There is less chance of a small number of data points influencing the shape of the function if those points that would otherwise make a large contribution to the sum are systematically deemphasized. If there is reason to suspect the presence of a false minimum, because of an unreasonably high sum of residuals or because of physically unreasonable values of some parameters, and there are enough data points to ensure a sufficient degree of overdetermination, the data can be divided randomly into two or more subsets, which can then be refined independently. The chances that all such subsets will have false minima in the same region of parameter space should be fairly small.

The fact that one of the ways false minima may be recognized is the existence of physically unreasonable values of parameters suggests another strategy for avoiding them, i.e., to build into the function, $f(\mathbf{x})$, for which we are seeking the minimum, criteria of physical reasonableness. This takes the form of a *penalty function* that adds to the value of $f(\mathbf{x})$ if the parameters differ from what are considered to be reasonable values. A technique widely used in crystal structure determination applies a penalty function if interatomic distances or bond angles differ markedly from those found in structures, or parts of structures, of similar type. The use of this type of penalty function is referred to as *restrained refinement*. It should not be confused with *constrained refinement*, which, as we shall see in Chapter 9, places much more severe restrictions on the model.

Chapter 7

Estimation of Uncertainty

When the minimum of a function of adjustable parameters has been found, we know the values of the adjustable parameters that give the "best fit" to the data. We are then faced with the question, "How confident can we be that the parameters that give the best fit correspond to the facts concerning the physical or chemical quantities we were trying to measure?" This question resolves itself into the problem of estimating two distinct quantities, which we call *precision* and *accuracy*.

Precision deals with the question of how closely it is possible to make a certain measurement with the available apparatus. Obviously, it is possible to measure the distance between two points much more precisely with a laser interferometer than with a meter stick. Precision can be estimated by observing how well, when we measure the same thing repeatedly or, more generally, measure it many times in different ways, the results agree with one another. It is thus a question that can be addressed using the methods of mathematical statistics. Accuracy, on the other hand, depends on the appropriateness and completeness of the model that is fitted to the data. The computer can adjust only those parameters that it is programmed to adjust. It has no way of knowing whether those parameters have physical meaning, and it has no way of knowing whether there are other physically meaningful parameters whose values, in a complete model, would be correlated with the adjusted parameters. The omission of meaningful parameters, and also the inclusion of meaningless ones, assumes values for them that may be different from the true ones, and leads to possibly precise, but nevertheless incorrect, values for the parameters in the model—an effect that is called *bias* by statisticians and *systematic error* by experimental scientists. Thus, good precision is a *necessary*, but *by no means sufficient*, condition for good accuracy.

We shall first discuss methods for estimating precision. We shall then consider what means may be used to judge whether a given level of precision can be confidently translated into a corresponding judgment of accuracy.

Estimates

The basic problem in estimation is to take a sample of observed data that are assumed to be drawn randomly from some underlying population and, by examining a suitably chosen function of these data points, to infer the values of the parameters of that population. Such a function of the data is known as a *statistic* or an *estimator*. If the expected value of an estimator, $f(x_i)$, is equal to the estimator itself, then the estimator is said to be *unbiased*.

$$\langle f(x_i)\rangle = \int_{-\infty}^{+\infty} f(x)\Phi(x)\,dx = \langle f(x)\rangle,$$

where $\Phi(x)$ is the probability density function of the underlying population distribution. Consider a sample of observations, x_i, drawn from a population represented by the function $\Phi(x)$, with mean μ and variance σ^2. The sample mean,

$$\bar{x} = (1/n)\sum_{i=1}^{n} x_i,$$

is an unbiased estimate of the population mean, μ, because

$$\langle\bar{x}\rangle = \langle(1/n)\sum_{i=1}^{n} x_i\rangle = (1/n)\sum_{i=1}^{n}\langle x_i\rangle,$$

and

$$\langle x_i\rangle = \int_{-\infty}^{+\infty} x\Phi(x)\,dx = \mu,$$

so that

$$\langle\bar{x}\rangle = (1/n)\sum_{i=1}^{n}\mu = \mu.$$

If an estimator is a linear function of a set of variables, x_i, of the form

$$f(\mathbf{x}) = \sum_{i=1}^{n} a_i x_i = \mathbf{a}^T\mathbf{x},$$

its variance is given by

$$\langle[f(\mathbf{x})]^2\rangle = \int_{-\infty}^{+\infty}[f(\mathbf{x})]^2\Phi(\mathbf{x})\,d\mathbf{x} = \int_{-\infty}^{+\infty}\Phi(\mathbf{x})\left[\sum_{i=1}^{n}\sum_{j=1}^{n} a_i a_j x_i x_j\right]d\mathbf{x},$$

or $\langle [f(\mathbf{x})]^2 \rangle = \mathbf{a}^T \mathbf{V} \mathbf{a}$, where

$$V_{ij} = \int_{-\infty}^{+\infty} x_i x_j \Phi(\mathbf{x})\, d\mathbf{x},$$

is the *variance-covariance matrix* of $\mathbf{x}$. In the case of a sample mean, for which $a_i = (1/n)$ for all i, and $V_{ij} = \sigma^2 \delta_{ij}$, since the x_i values are *independent* observations drawn from the same population, we obtain

$$\left\langle (\bar{x} - \mu)^2 \right\rangle = (1/n^2) \sum_{i=1}^{n} \sigma^2 = \sigma^2 / n.$$

Thus, the variance of the estimated mean of the population is the variance of the population divided by the number of observations. It can, therefore, be made arbitrarily small by taking a large enough number of observations.

To estimate the population variance we need to find $\langle (x_i - \mu)^2 \rangle$. We may start with the identity

$$\sum_{i=1}^{n} \left\langle (x_i - \bar{x})^2 \right\rangle \equiv \sum_{i=1}^{n} \left[\langle x_i^2 \rangle - 2 \langle x_i \bar{x} \rangle \right] + n \langle \bar{x}^2 \rangle.$$

Evaluating the terms on the right individually, we have

$$\langle x_i^2 \rangle = \left\langle \left[(x_i - \mu) + \mu \right]^2 \right\rangle = \left\langle (x_i - \mu)^2 \right\rangle + 2 \langle \mu (x_i - \mu) \rangle + \langle \mu^2 \rangle.$$

Since $\langle x_i \rangle = \mu$, the middle term vanishes, while $\langle (x_i - \mu)^2 \rangle = \sigma^2$, so that $\langle x_i^2 \rangle = \sigma^2 + \mu^2$. By a similar procedure, since $\langle (\bar{x} - \mu)^2 \rangle = \sigma^2 / n$, $\langle \bar{x}^2 \rangle = \sigma^2 / n + \mu^2$. Finally,

$$\langle x_i \bar{x} \rangle = \left\langle (1/n) \sum_{j=1}^{n} x_i x_j \right\rangle = (1/n) \left[\sigma^2 + \mu^2 + (n-1) \mu^2 \right] = \langle \bar{x}^2 \rangle.$$

Therefore

$$\sum_{i=1}^{n} \left\langle (x_i - \bar{x})^2 \right\rangle = n(\sigma^2 + \mu^2) - \sigma^2 - n\mu^2 = (n-1)\sigma^2,$$

and

$$s^2 = \left[1/(n-1) \right] \sum_{i=1}^{n} (x_i - \bar{x})^2$$

is an unbiased estimate of the population variance, σ^2.

Although the expression

$$\sum_{i=1}^{n} (x_i - \bar{x})^2$$

would seem to imply that it is first necessary to compute the mean, $\bar{x}$, then to form the individual differences, and finally to compute the sum of squares, in practice statisticians never perform the computation in this way. Instead, they make use of an algebraic identity, as follows:

$$\sum_{i=1}^{n} (x_i - \bar{x})^2 \equiv \sum_{i=1}^{n} (x_i^2 - 2x_i\bar{x} + \bar{x}^2),$$

but

$$\bar{x} = (1/n) \sum_{i=1}^{n} x_i,$$

which, on substitution and simplification, gives

$$\sum_{i=1}^{n} (x_i - \bar{x})^2 \equiv \sum_{i=1}^{n} x_i^2 - (1/n)\left(\sum_{i=1}^{n} x_i\right)^2.$$

Thus, the procedure is to sum, in parallel, the first powers and the squares, giving two sums, S_1, and S_2. When this is complete, the sample mean is $\bar{x} = S_1/n$, and the sample variance is $s^2 = [1/(n-1)](S_2 - S_1^2/n)$. The sum S_2 is usually called the *raw sum of squares* or the *uncorrected sum of squares*. The quantity $S_2 - S_1^2/n$ is called the *corrected sum of squares*, although *sum of squared residuals* would seem to be a less confusing term. Nothing is being "corrected;" rather, a new statistic is being computed. The quantity S_1^2/n is, not infrequently in the statistical literature, called the *correction factor*, although it is a *term*, not a *factor*. Another booby trap the statisticians set for the unwary experimenter is to call the quantity $(n-1)$, or $(n-p)$ in the case of a multiple parameter fit, the number of *degrees of freedom*, meaning not, as in physics, the number of ways a model can vary, but the number of independent pieces of information going into the computation. The sum by which the variance is computed contains n terms, but only $(n-1)$ of them are independent, because of the constraint

$$\sum_{i=1}^{n} (x_i - \bar{x}) = 0.$$

Thus far we have considered only samples drawn from the same population, with the same mean and the same variance. Let us now suppose that we have a number of different measurements of the same quantity, but the measurements have, for some reason, been made at different levels of precision, so that the populations from which they are drawn have the same mean but different variances. Clearly, each measurement, x_i, is an unbiased estimate of the common mean, and any linear function of the measurements,

$$\bar{x} = \sum_{i=1}^{n} a_i x_i,$$

is also an unbiased estimate of the mean, provided only that

$$\sum_{i=1}^{n} a_i = 1.$$

We would like, however, to choose a function of the measurements that gives the most precise estimate of the common, population mean, i.e., the estimate with the smallest variance. The variance of our linear function is

$$\left\langle (\bar{x} - \mu)^2 \right\rangle = \langle \bar{x}^2 \rangle - \mu^2 = \int_{-\infty}^{+\infty} \left[\sum_{i=1}^{n} a_i x_i \right]^2 \Phi(x)\, dx - \mu^2,$$

and, again, because the individual measurements are independent,

$$\int_{-\infty}^{+\infty} x_i x_j \Phi(x)\, dx = 0 \qquad \text{if } i \neq j,$$

so that

$$\langle \bar{x}^2 \rangle = \sum_{i=1}^{n} a_i^2 \sigma_i^2 + \mu^2.$$

Because of the constraint that the sum of the a_i values must be 1, only $n - 1$ of them are independent, and

$$a_n = 1 - \sum_{i=1}^{n-1} a_i.$$

For minimum variance the partial derivatives of $\langle \bar{x}^2 \rangle$ with respect to all a_i values must vanish, giving $\partial \langle \bar{x}^2 \rangle / \partial a_i = 2a_i\sigma_i^2 - 2a_n\sigma_n^2 = 0$, or $a_i\sigma_i^2 = a_n\sigma_n^2 = k$, for $i = 1, 2, 3, \ldots, n - 1$, and $a_i = k/\sigma_i^2$. Now

$$\sum_{i=1}^{n} a_i = \sum_{i=1}^{n} k/\sigma_i^2 = k \sum_{i=1}^{n} 1/\sigma_i^2 = 1,$$

so that

$$k = 1 \Big/ \sum_{i=1}^{n} (1/\sigma_i^2).$$

If we define the *weight* as $w = 1/\sigma_i^2$, the *weighted mean* with minimum variance is

$$\bar{x} = \sum_{i=1}^{n} w_i x_i \Big/ \sum_{i=1}^{n} w_i.$$

Its variance is given by

$$\left\langle (\bar{x} - \mu)^2 \right\rangle = \sum_{i=1}^{n} w_i^2 \sigma_i^2 \Big/ \left[\sum_{i=1}^{n} w_i \right]^2 = 1 \Big/ \sum_{i=1}^{n} w_i.$$

The weighted mean, defined in this way, is identical to the least squares estimate of μ, which can be seen as follows: We seek the value, $\hat{\mu}$, that minimizes the function

$$f(\mu) = \sum_{i=1}^{n} w_i (x_i - \mu)^2.$$

At the minimum, the derivative with respect to μ vanishes.

$$\frac{df(\mu)}{d\mu} = -2 \sum_{i=1}^{n} w_i (x_i - \mu) = 0,$$

which leads directly to

$$\hat{\mu} = \sum_{i=1}^{n} w_i x_i \Big/ \sum_{i=1}^{n} w_i = \bar{x}.$$

Thus, the weighted least squares estimate is identical to the minimum variance linear estimate of the population mean.

We have seen that, if a set of observations is drawn from a single population, we can estimate the mean and the variance of the population, and that, if the observations come from different populations with the same mean but with different, known variances, we can estimate the mean with minimum variance. But what if we do not know the variances? If we have some way of knowing, estimating, or guessing the *relative* variances, so that weights can be assigned according to $w_i = k/\sigma_i^2$, we can estimate k by noting that, in computing the weighted mean, each term in the sum has been scaled so that all terms, $(x_i - \bar{x})^2$ are expressed as fractions of their individual population variances, σ_i^2. This suggests that, if

$$s^2 = [1/(n-1)] \sum_{i=1}^{n} (x_i - \bar{x})^2$$

is an estimate of σ^2, that the quantity

$$[1/(n-1)] \sum_{i=1}^{n} (x_i - \bar{x})^2 / \sigma_i^2$$

should be approximately equal to 1, and that

$$[1/(n-1)] \sum_{i=1}^{n} w_i (x_i - \bar{x})^2$$

should therefore be an estimate of k. A great deal of data analysis in experimental science is based on arguments of this sort; they are often well-justified, particularly if there is good reason to believe that the model is correct, but they are sometimes questionable.

The Precision of Estimates of Precision

All of the discussion of estimating mean and variance so far has been independent of the shape of the probability density function, $\Phi(x)$, of the underlying population from which the observations, x_i, were drawn. Thus, an unbiased estimate of the mean, μ, may be computed without reference to the shape of the underlying distribution function. Likewise, the precision with which we have estimated the mean, σ^2, can be estimated without reference to the shape of the distribution function. If we wish to know how confident we can be in our estimate of the variance, however, we must assume a functional form for the underlying population distribution. In practice by far the most common assumption made is that the underlying population has a normal, or Gaussian, distribution, defined by

$$G(x, \mu, \sigma) = (2\pi\sigma^2)^{-1/2}\exp\{-(1/2)[(x-\mu)/\sigma]^2\}.$$

There are two common justifications for this assumption. The first is a result known as the *central limit theorem*, which we shall, following the custom of most elementary and many more advanced books on statistics, state without proof: Given a set of n random variables, x_i, drawn from populations with means μ_i and variances σ_i^2, the function

$$y = \sum_{i=1}^{n} a_i x_i$$

has a probability density function that tends, as n becomes very large, to a normal density function with mean

$$\mu_y = \sum_{i=1}^{n} a_i \mu_i$$

and variance

$$\sigma_y^2 = \sum_{i=1}^{n} a_i^2 \sigma_i^2.$$

The second justification is simply the practical experience that tests based on the normal distribution function give reasonable results for most bell-shaped curves, provided the area in the tails is not too large a fraction of the total.

We have already seen that, if n observations, x_i, are drawn from a population with mean μ and variance σ^2 the quantity

$$s^2 = (1/\nu)\sum_{i=1}^{n}(x_i - x)^2,$$

where $\nu = (n-1)$ is the number of degrees of freedom, is an unbiased estimate of σ^2. If the population from which the observations are drawn has

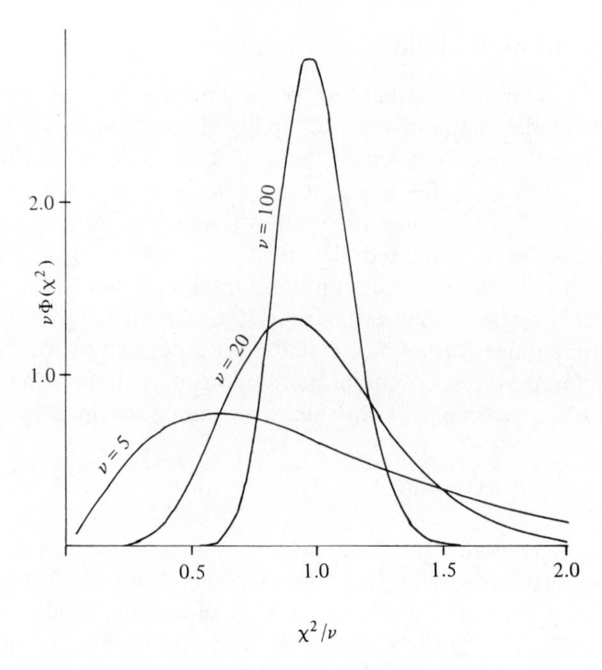

Fig. 7.1. The χ^2 distribution for several values of ν.

a normal distribution, the quantity $\chi^2 = \nu s^2/\sigma^2$ has a density function

$$\Phi(\chi^2) = \left[2^{\nu/2}\Gamma(\nu/2)\right]^{-1}\exp(-\chi^2/2)(\chi^2)^{\nu/2-1} \quad \text{for } \chi^2 \geqslant 0,$$

$$\Phi(\chi^2) = 0 \quad \text{otherwise.}$$

$\Gamma(\nu/2)$ is the *gamma function*, whose properties are discussed in Appendix D. The function $\Phi(\chi^2)$ is the *chi-squared distribution with* ν *degrees of freedom*.[1] Figure 7.1 shows the χ^2 distribution for several values of ν. It is a skewed, bell-shaped curve that gets sharper as ν increases.

Models with More than One Parameter

The estimation of the mean of a population from a sample of observations drawn from that population can be viewed, alternatively, as the fitting to

[1] In most books on statistics a table at the end gives the values of χ^2 for which the cumulative distribution function,

$$\Psi(\chi^2) = \int_0^{\chi^2}\Phi(t)\,dt,$$

is equal to various numbers between zero and one for various choices of ν. In crystallography the values of ν are often large and do not appear in the table. For this reason I have included in Appendix G a FORTRAN program for computing the cumulative chi-squared distribution for any cutoff point and any number of degrees of freedom.

the data of a model with one parameter. That is, $x_i = \mu + \epsilon_i$, where ϵ_i represents a set of random variables drawn from a population with zero mean. We have seen that the sample mean, $\bar{x}$, and the least squares estimate, $\hat{\mu}$, are identical, minimum-variance, unbiased estimates of the parameter, μ. We shall now consider a linear model with two parameters, largely because such a model illustrates, with easily tractable mathematics, a number of principles that carry over into an arbitrarily large number of parameters.

We start with a set of observations, y_i, taken under various conditions that represent different values of an adjustable parameter, x, which we assume relates to the values of y_i by a model of the form $y_i = a + bx_i + \epsilon_i$, where ϵ_i again represents a set of random variables drawn from a population with zero mean. We wish to estimate the "best" values of the parameters, a and b, and their variances. As before, we wish to find the values, $\hat{a}$ and $\hat{b}$, of the parameters that minimize the function

$$f(a,b) = \sum_{i=1}^{n} (y_i - a - bx_i)^2.$$

At the minimum the partial derivatives vanish, so we have

$$\frac{\partial f}{\partial a} = -2\sum_{i=1}^{n} (y_i - a - bx_i) = 0,$$

and

$$\frac{\partial f}{\partial b} = -2\sum_{i=1}^{n} x_i(y_i - a - bx_i) = 0.$$

Rearranging terms, we obtain the pair of simultaneous equations

$$na + b\sum_{i=1}^{n} x_i = \sum_{i=1}^{n} y_i,$$

and

$$a\sum_{i=1}^{n} x_i + b\sum_{i=1}^{n} x_i^2 = \sum_{i=1}^{n} x_i y_i.$$

If we designate by $\mathbf{v}$ the column vector whose elements are a and b, define the *design matrix*, $\mathbf{A}$, by $A_{i1} = 1$ and $A_{i2} = x_i$, and designate by $\mathbf{y}$ the vector formed by the y_i values, these equations become $(\mathbf{A}^T\mathbf{A})\mathbf{v} = \mathbf{A}^T\mathbf{y}$, which has the solution, $\mathbf{v} = (\mathbf{A}^T\mathbf{A})^{-1}\mathbf{A}^T\mathbf{y}$. Now

$$(\mathbf{A}^T\mathbf{A})^{-1} = \begin{bmatrix} \sum_{i=1}^{n} x_i^2/\Delta & -\sum_{i=1}^{n} x_i/\Delta \\ -\sum_{i=1}^{n} x_i/\Delta & n/\Delta \end{bmatrix},$$

where $\Delta = n\sum x_i^2 - (\sum x_i)^2$, and $(\mathbf{A}^T\mathbf{A})^{-1}\mathbf{A}^T$ is a matrix with two rows and n columns. The elements of the first row are

$$\left[(\mathbf{A}^T\mathbf{A})^{-1}\mathbf{A}^T\right]_{1j} = \left(\sum_{i=1}^{n} x_i^2 - x_j \sum_{i=1}^{n} x_i\right)\Big/\Delta = \alpha_j.$$

The elements of the second row are

$$\left[(\mathbf{A}^T\mathbf{A})^{-1}\mathbf{A}^T\right]_{2j} = \left(nx_j - \sum_{i=1}^{n} x_i\right)\Big/\Delta = \beta_j.$$

Thus $\hat{a} = \sum_{i=1}^{n}\alpha_i y_i$, and $\hat{b} = \sum_{i=1}^{n}\beta_i y_i$, and, therefore, $\hat{a}$ and $\hat{b}$ are both linear functions of the independent, random variables, y_i, drawn from a population with variance σ^2. The variances of $\hat{a}$ and $\hat{b}$ are then

$$\langle(\hat{a} - a)^2\rangle = \sigma^2 \sum_{i=1}^{n} \alpha_i^2,$$

and

$$\langle(\hat{b} - b)^2\rangle = \sigma^2 \sum_{i=1}^{n} \beta_i^2.$$

Straightforward algebra shows that

$$\langle(\hat{a} - a)^2\rangle = \sigma^2 \sum_{i=1}^{n} x_i^2/\Delta = \sigma^2(\mathbf{A}^T\mathbf{A})_{i1}^{-1},$$

and, similarly, that $\langle(\hat{b} - b)^2\rangle = \sigma^2(\mathbf{A}^T\mathbf{A})_{22}^{-1}$. Further, the covariance of $\hat{a}$ and $\hat{b}$ is $\langle(\hat{a} - a)(\hat{b} - b)\rangle = \sigma^2(\mathbf{A}^T\mathbf{A})_{12}^{-1}$. Thus, the variance-covariance matrix of $\hat{a}$ and $\hat{b}$ is the inverse of the least squares Hessian matrix multiplied by the variance of the population from which the observations are drawn.

Just as in the case of the "one-parameter fit," the quantity

$$s^2 = \left[1/(n-1)\right] \sum_{i=1}^{n} (x_i - \bar{x})^2$$

is an unbiased estimate of σ^2, the variance of the population from which the observations, x_i, were drawn, in the two-parameter fit σ^2 may be estimated by the quantity

$$s^2 = \left[1/(n-2)\right] \sum_{i=1}^{n} (y_i - \hat{y}_i)^2,$$

where $\hat{y}_i = \hat{a} + \hat{b}x_i$ is the *predicted* value of y_i for $x = x_i$. The number of degrees of freedom is now $(n - 2)$, because the fitting process causes two of the residuals to be dependent on all of the others. In general, if there are p parameters, there are $(n - p)$ degrees of freedom.

In our least squares example (page 13) we saw that the inverse of the Hessian matrix was

$$\mathbf{H}^{-1} = \begin{bmatrix} 0.50000 & -0.20000 & 0.01600 \\ -0.20000 & 0.13600 & -0.01280 \\ 0.01600 & -0.01280 & 0.00128 \end{bmatrix}.$$

The sum of squared residuals is 0.2462, *and the number of degrees of freedom is* $10 - 3 = 7$. *The estimated overall variance of the population from which the data are drawn is therefore* $0.2462/7 = 0.03517$, *and the variances of the estimated parameters are*

$$\left\langle (b_0 - \hat{b}_0)^2 \right\rangle = 0.0176,$$

$$\left\langle (b_1 - \hat{b}_1)^2 \right\rangle = 0.00478, \quad \text{and}$$

$$\left\langle (b_2 - \hat{b}_2)^2 \right\rangle = 0.000045.$$

The parameters are therefore $b_0 = 0.567(133)$, $b_1 = 0.218(69)$, *and* $b_2 = -0.0034(67)$, *where the numbers in parentheses designate the standard deviations of the corresponding parameters, with the understanding that these digits multiply the same power of* 10 *as the least significant digits given for the parameter. That is,* $-0.0034(67)$ *means that the standard deviation is* 0.0067. *We may conclude that* b_2 *is probably not "significantly" different from zero. We shall discuss significance in more detail in the next chapter.*

Estimates of Uncertainty When the Algorithm Is Not Least Squares

We saw earlier that the method of least squares, although it gives the best linear unbiased estimate of the model parameters irrespective of the distribution of errors in the observations, yields the maximum likelihood estimate only if the error distribution is normal, or Gaussian. We also saw that alternative algorithms could be expected to give "better," meaning lower variance, estimates if the error distribution were not normal, or if the variances of the individual observations were not reliably known, so that the various observations could be properly weighted. Much of the preceding discussion of uncertainty estimates has assumed that the error distribution was Gaussian with (at least approximately) known variances, partly because of the implications of the central limit theorem, but also because the Gaussian distribution can be (or, at any rate, has been) analyzed reasonably rigorously. Analysis of the more robust/resistant variants of least squares involves much more intractable mathematics: assumptions that, inherently, cannot be rigorously justified, and integrals that can be evaluated only numerically. For this reason most existing analysis of uncertainty estimates related to the robust/resistant algorithms is based on determining the bias that would be introduced into the estimate of variance for a normal distribution by the variant procedure, and correcting for it by multiplying the biased estimate of variance by a constant factor.

We have seen that the function

$$s^2 = \sum_{i=1}^{n} (y_i - \hat{y}_i)^2/(n-p)$$

is an estimator of the population variance. It is evident that a small number of observations, y_i, that are very different from their expected values, $\hat{y}_i$, have a very large influence on the value of this function. On the other hand if the error distribution is normal, the quantity $(|R|_m/0.6745)^2$, where $|R|_m$ represents the median absolute value of $[(y_i - \hat{y}_i)/\sigma_i]$, is also an unbiased estimate of σ^2 for the population of errors in the data. The difference between the two estimates is that the latter one is totally unaffected by all residuals whose absolute values are larger than the median. A data set could consist of a sample nearly half of whose members were gross blunders, and the other half (plus a little bit) would be faithfully represented by this estimator. In other words, it is extremely resistant.

One drawback to using $|R|_m$ as an estimator for the variance of the distribution is that it is not a function of all of the data, so that it cannot be computed concurrently with the sum-of-squares, the gradient, and the elements of the Hessian matrix. Instead it is necessary to store all of the individual values of the residuals and, at the end of a cycle, to apply a partial sorting algorithm to the resulting list to find which member is the median value. An alternative estimator, which is a function of all of the data, but which is also resistant, is

$$\hat{\sigma}^2 = [\beta/(n-p)] \sum_{i=1}^{n} [R_i\phi(R_i/S)]^2,$$

where $\phi(R_i/S)$ is a robust/resistant weighting function, S is a scale factor computed on a previous cycle, and

$$1/\beta = \int_{-\infty}^{+\infty} [x\phi(x/S)]^2 \exp(-x^2/2)\, dx$$

is the expected value of $[x\phi(x/S)]^2$ if x is a normally distributed random variable. This estimate of the variance will be unbiased if the data are drawn mostly from a normal distribution, with a small number of outliers, but will tend to be biased low if the real distribution is substantially longer-tailed than a normal distribution.

Chapter 8

Significance and Accuracy

The F Distribution

A question that is encountered very frequently in the analysis of experimental data is, "Does the value of an observation, y_i, really depend on a parameter, x?" In the two-parameter problem of the previous chapter the question could be phrased, "Is the value of the coefficient, b, 'significantly' different from zero?" One way to answer this question is to find some *independent* estimate of σ^2 and determine whether the spread of the values of y_i about their mean, $\bar{y}$, is different from the spread of the observed values of y_i, about their predicted values, $\hat{y}_i$. Suppose, for example, that for each value of x we measure y several times. If we designate by m_i the number of times we measure y when $x = x_i$, the function we wish to minimize becomes

$$f(a,b) = \sum_{i=1}^{n} \sum_{j=1}^{m_i} (y_{ij} - a - bx_i)^2.$$

Designating by $\bar{y}_i$ the mean of the values of y when $x = x_i$, we can rewrite that expression

$$f(a,b) = \sum_{i=1}^{n} \sum_{j=1}^{m_i} \left[(y_{ij} - \bar{y}_i) + (\bar{y}_i - a - bx_i)\right]^2,$$

which, when expanded, becomes

$$f(a,b) = \sum_{i=1}^{n} \sum_{j=1}^{m_i} \Big[(y_{ij} - \bar{y}_i)^2 + (\bar{y}_i - a - bx_i)^2$$
$$+ 2(y_{ij} - \bar{y}_i)(\bar{y}_i - a - bx_i)\Big].$$

The third term within the square brackets can be written

$$2(\bar{y}_{ij} - a - bx_i)\left[\sum_{j=1}^{m_i} y_{ij} - m_i\bar{y}_i\right],$$

but

$$\sum_{j=1}^{m_i} y_{ij} = m_i\bar{y}_i,$$

so the factor in square brackets vanishes, and

$$f(a,b) = \sum_{i=1}^{n}\left[m_i(\bar{y}_i - a - bx_i)^2 + \sum_{j=1}^{m_i}(y_{ij} - \bar{y}_i)^2\right].$$

This expression has two sets of terms that may be viewed as a sum of squares resulting from the difference between the model and the mean value of y at $x = x_i$ and a sum of squares resulting from the spread of the individual values of y about their means.

Each term of the type $\sum_{j=1}^{m_i}(y_{ij} - \bar{y}_i)^2$ is an independent, and unbiased, estimate of the quantity $(m_i - 1)\sigma^2$, with $(m_i - 1)$ degrees of freedom. The expression

$$s_R^2 = \sum_{i=1}^{n}\sum_{j=1}^{m_i}(y_{ij} - \bar{y}_i)^2,$$

known as the *replication sum of squares*, is therefore an estimate of the quantity $\sigma^2\sum_{i=1}^{n}(m_i - 1)[= \sum_{i=1}^{n} m_i - n]$ and, if the y_i values are normally and independently distributed, has the χ^2 distribution with $(\sum_{i=1}^{n} m_i - n)$ degrees of freedom. Similarly, the quantity

$$s_L^2 = \sum_{i=1}^{n} m_i(\bar{y}_i - a - bx_i)^2,$$

known as the *lack-of-fit sum of squares*, is an unbiased estimate of $(n - 2)\cdot\sigma^2$, and has a χ^2 distribution with $(n - 2)$ degrees of freedom. The ratio of these independent estimates of σ^2, $F = [s_L^2/(n - 2)]/[s_R^2/(\sum_{i=1}^{n} m_i - n)]$, has the very important *F distribution* with $(n - 2)$ and $(\sum_{i=1}^{n} m_i - n)$ degrees of freedom. The F distribution function for ν_1 degrees of freedom in the numerator and ν_2 degrees of freedom in the denominator is

$$\Phi(F,\nu_1,\nu_2) = \frac{\Gamma[(\nu_1 + \nu_2)/2]}{\Gamma(\nu_1/2)\Gamma(\nu_2/2)}\left(\frac{\nu_1}{\nu_2}\right)^{\nu_1/2} \times \frac{F^{(\nu_1-2)/2}}{[1 + (\nu_1/\nu_2)F]^{(\nu_1+\nu_2)/2}} \quad \text{for } F \geqslant 0$$

$$\Phi(F,\nu_1,\nu_2) = 0 \quad \text{otherwise.}$$

Figure 8.1 shows the F density function for several pairs of values of ν_1 and ν_2. It is a curve with a maximum value in the vicinity of 1 and, if ν_1 is at least 2, falls to zero at $F=0$ and $F \to \infty$. A FORTRAN language function given in Appendix G enables the computation of the probability that the value of F will be less than or greater than specified values if the two estimates of σ^2 are consistent. If the model gives an adequate fit to the data, the F ratio should be near 1. If the ratio exceeds the 95% point of the cumulative F distribution, there is only a 5% chance that the model is adequate.

Each of the data points in our least squares example is the mean of five repeated observations at each value of x. The raw data are

i	x_i	y_{i1}	y_{i2}	y_{i3}	y_{i4}	y_{i5}	$\bar{y}_i$
1	0.061	−0.073	−0.060	0.831	0.742	0.338	0.355
2	0.545	1.222	0.697	0.980	0.783	0.789	0.894
3	1.464	0.734	1.216	0.644	0.951	0.801	0.869
4	2.730	1.645	1.572	1.220	0.827	0.443	1.141
5	4.218	1.078	1.500	1.399	1.938	2.074	1.598
6	5.782	1.600	1.442	2.102	2.171	1.507	1.764
7	7.270	1.394	1.684	1.694	1.594	2.064	1.686
8	8.536	1.967	2.652	1.347	1.836	2.652	2.091
9	9.455	2.143	3.049	2.386	2.123	2.759	2.492
10	9.938	2.548	3.124	2.248	1.573	2.707	2.440

The replication sum of squares is 6.771, *the number of degrees of freedom is* $50-10=40$, *so* 0.1693 *is an estimate of the variance of the population from which the data are drawn. If we test the model that all of these observations are drawn from a population with the same mean, we have a one parameter fit with* $b_0 = 1.153$. *The lack-of-fit sum of squares is* 22.62 *with* $10-1=9$ *degrees of freedom, for an F ratio of* 14.84. *The cumulative F distribution function is essentially* 1.0 *for this value of F, so we may reject the hypothesis that all of the observations have the same mean with an exceedingly high confidence. On the other hand, if we try a two-parameter fit, we find* $b_0 = 0.609$, $b_1 = 0.185$, *and the lack-of-fit sum of squares is* 1.275 *with* $10-2=8$ *degrees of freedom. The F ratio is* 0.941. *The cumulative F distribution function is* 0.506, *so we conclude that the linear model is a perfectly adequate fit to the data, and that the quadratic term is not significantly different from zero.*

Returning to the question of whether the parameter, b, is "significantly" different from zero, we may *constrain* it to be equal to zero and calculate the value, a, that minimizes the *constrained sum of squares*

$$s_c^2 = \sum_{i=1}^{n} (y_i - a)^2.$$

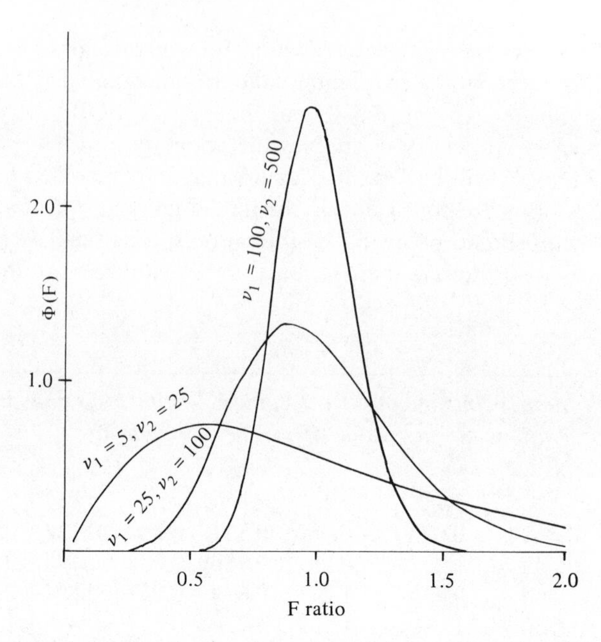

Fig. 8.1. The F distribution for several pairs of values of ν_1 and ν_2.

If the constrained model is adequate, $s_c^2/(n-1)$ is an estimate of σ^2 with $(n-1)$ degrees of freedom. If b has *any* mathematical influence on the model, s_c^2 will be somewhat larger than the unconstrained sum of squares,

$$s_u^2 = \sum_{i=1}^{n} (y_i - a - bx_i)^2,$$

but $s_\mu^2/(n-2)$ is an estimate of σ^2 with only $(n-2)$ degrees of freedom. If $s_L^2 = (n-1)\sigma^2$, and $s_u^2 = (n-2)\sigma^2$, then $s_c^2 - s_u^2 = [(n-1)-(n-2)]\sigma^2$, and therefore the difference between the constrained and unconstrained sums of squares is an estimate of σ^2 with one degree of freedom. The F ratio $F = (s_c^2 - s_u^2)/[s_u^2/(n-1)]$ has the F distribution with 1 and $(n-1)$ degrees of freedom. If the ratio exceeds the 95% point of the appropriate cumulative F distribution, there is a 0.95 probability that the parameter, b, is different from zero.

In our example, the one-parameter, constrained sum of squares is 4.524. *The two-parameter, unconstrained sum of squares is* 0.2551, *with* 8 *degrees of freedom. The F ratio is* $(4.524 - 0.255)/(0.255/8) = 133.86$. *The cumulative F distribution function for* 1 *and* 8 *degrees of freedom is* 0.999997, *so the second parameter,* b_1, *is highly significant. The three-parameter, unconstrained sum of squares is* 0.2462, *for an F ratio of* $(0.2551 - 0.2462)/(0.2462/7) = 0.2523$. *The cumulative F distribution function with* 1 *and* 7 *degrees of freedom is* 0.3691, *so the quadratic term, in agreement with our other tests, is not significant.*

In the more general case where we wish to compare one model with p_1 parameters with another model with a larger number, p_2, of parameters, the F ratio is $F = [(s_1^2 - s_2^2)/(p_2 - p_1)]/[s_2^2/(n - p_2)]$, and it has the F distribution with $(p_2 - p_1)$ and $(n - p_2)$ degrees of freedom.

It should be emphasized that an F ratio appreciably greater than 1 is a *necessary* condition for the unconstrained model to be a "significantly" better fit to the data than the constrained one. It is a *sufficient* condition, however, only in a blindly mathematical sense. An unconstrained model that violates the laws of nature cannot be considered a better model, no matter what the value of the F ratio may be. But an F ratio that is *not* appreciably greater than 1 indicates that the data do not contain sufficient information to distinguish between the two models.

Student's *t* Distribution

An important special case of a "constrained" fit is the one of determining whether a population mean, μ, is significantly different from some hypothetical value, μ_0. This may be treated as the comparison of a one-parameter fit with a zero-parameter fit. The difference of sums of squares in this case is

$$(s_c^2 - s_u^2) = \sum_{i=1}^{n} (x - \mu_0)^2 - \sum_{i=1}^{n} (x_i - \bar{x})^2.$$

Expanding this expression, we get

$$(s_c^2 - s_u^2) = \sum_{i=1}^{n} (x_i^2 - 2x_i\mu_0 + \mu_0^2 - x_i^2 + 2x_i\bar{x} - \bar{x}^2).$$

which simplifies to

$$(s_c^2 - s_u^2) = n(\mu_0^2 - \bar{x}^2) + 2(\bar{x} - \mu_0)\sum_{i=1}^{n} x_i.$$

Now $\sum_{i=1}^{n} x_i = n\bar{x}$, so that $(s_c^2 - s_u^2) = n(\mu_0^2 - \bar{x}^2) + 2n(\bar{x}^2 - \bar{x}\mu_0) = n(\bar{x} - \mu_0)^2$. The F ratio is then $F = [n(\bar{x} - \mu_0)^2]/[s_u^2/(n-1)]$ and has the F distribution with 1 and $(n - 1)$ degrees of freedom. Consider the quantity

$$t = \pm F^{1/2} = \left[n^{1/2}(\bar{x} - \mu_0)\right]/\left[s_u^2/(n-1)\right]^{1/2}.$$

Substituting t^2 for F in $\Phi(F, 1, \nu)$, we obtain

$$\Phi(t) = \frac{\Gamma[(\nu + 1)/2]}{(\nu\pi)^{1/2}\Gamma(\nu/2)}\left[1 + (t^2/\nu)\right]^{-(\nu+1)/2}.$$

This probability density function was introduced by W. S. Gossett, who published under the pseudonym "Student," and is known as *Student's t* *distribution with ν degrees of freedom*. Although it is actually a special case of the F distribution,[1] its importance in the establishment of confidence limits for means is so great that it is treated and tabulated as a special distribution. A FORTRAN language function for computing the cumulative t distribution function is given in Appendix G.

Correlation

We have already seen that the inverse of the Hessian matrix of a sum-of-squares function is the variance-covariance matrix of the estimated parameters. In our two-parameter example the Hessian matrix was

$$\mathbf{H} = \mathbf{A}^T\mathbf{A} = \begin{bmatrix} n & \sum_{i=1}^{n} x_i \\ \sum_{i=1}^{n} x_i & \sum_{i=1}^{n} x_i^2 \end{bmatrix}.$$

The controlled parameter, x_i, is assumed to be measured from some arbitrary origin, x_0, so that the off-diagonal elements of the Hessian matrix are, in general, nonzero, and the off-diagonal elements of the variance-covariance matrix are also nonzero. The quantity $\rho_{12} = H_{12}^{-1}/(H_{11}^{-1}H_{22}^{-1})^{1/2}$ is known as the *correlation coefficient* for variables 1 and 2. In a general, multiparameter problem the *correlation matrix*, $\boldsymbol{\rho}$, is defined by

$$\rho_{ij} = H_{ij}^{-1}/\left(H_{ii}^{-1}H_{jj}^{-1}\right)^{1/2}.$$

If the design matrix, $\mathbf{A}$, has at least as many rows as columns, and the columns are linearly independent, $\mathbf{H}$ and $\mathbf{H}^{-1}$ are positive definite, and by Schwarz's inequality, $H_{ij}^2 \leqslant H_{ii}H_{jj}$. Thus ρ_{ij}, for $i \neq j$, is a number in the region $-1 \leqslant \rho_{ij} \leqslant 1$. (It should be noted that the existence of correlation coefficients whose values are close to ± 1 does not preclude the correct computation of the variances of the individual parameters; a diagonal element of the variance-covariance matrix is the correct variance for the marginal distribution of the corresponding variable. However, if the Hessian matrix does not contain rows and columns corresponding to *all* variables that have correlations with the variable of interest, the computed variance will be too small, and the apparent precision will be illusory.)

Although the presence of high correlation presents no *mathematical* problems, it may present practical, computational problems. If $H_{ij}^2 = H_{ii}H_{jj}$ for any unequal pair i and j, then $\mathbf{H}$ is singular, and has no inverse. If, however, H_{ij}^2 is nearly equal to $H_{ii}H_{jj}$, the computation of the inverse involves small differences between much larger numbers; the problem is

[1]Note that although the F distribution is defined only for $F \geqslant 0$, the t distribution is defined for all real t.

said to be *ill-conditioned.* Because all computers have finite word lengths, arithmetic precision may be lost, and the results may be unstable or even entirely erroneous. It would seem to be desirable, therefore, to construct models in which correlation is reduced to as low a level as possible.

In our two-parameter example let us suppose that we shift the origin to the mean of all the values of the independent parameter, x_i, so that the model becomes $y_i = a + b(x_i - \bar{x}) + \epsilon_i$. The Hessian matrix then becomes

$$\mathbf{H} = \begin{bmatrix} n & \sum_{i=1}^{n}(x_i - \bar{x}) \\ \sum_{i=1}^{n}(x_i - \bar{x}) & \sum_{i=1}^{n}(x_i - \bar{x})^2 \end{bmatrix} = \begin{bmatrix} n & 0 \\ 0 & \sum_{i=1}^{n}(x_i - \bar{x})^2 \end{bmatrix}.$$

Thus, by a shift of the origin, the correlation between a and b has disappeared.

For another example, suppose that there is apparent curvature in the function that describes y in terms of x, so that the model must be a quadratic one of the form $y_i = a + bx_i + cx_i^2 + \epsilon_i$. We may assume, without any loss of generality, that the origin has been chosen so that $\bar{x} = 0$, and we are free to choose the points, x_i, in a symmetric way with respect to the origin so that $\sum_{i=1}^{n} x_i^3 = 0$. We may further assume that we have scaled the x axis so that all values of x_i fall in the range $-1 \leqslant x_i \leqslant 1$. Nevertheless we find that the Hessian matrix is

$$\mathbf{H} = \begin{bmatrix} n & 0 & \sum_{i=1}^{n} x_i^2 \\ 0 & \sum_{i=1}^{n} x_i^2 & 0 \\ \sum_{i=1}^{n} x_i^2 & 0 & \sum_{i=1}^{n} x_i^4 \end{bmatrix},$$

so that, despite all our maneuvering, we have not been able to avoid a correlation between the constant term and the coefficient of the quadratic term.

Suppose, however, that instead of a power series, we use for our model the function $y_i = a + bx_i + c(2x_i^2 - 1) + \epsilon_i$. The Hessian matrix then is

$$\mathbf{H} = \begin{bmatrix} n & 0 & \sum_{i=1}^{n}(2x_i^2 - 1) \\ 0 & \sum_{i=1}^{n} x_i^2 & 0 \\ \sum_{i=1}^{n}(2x_i^2 - 1) & 0 & \sum_{i=1}^{n}(2x_i^2 - 1)^2 \end{bmatrix},$$

and it is not difficult to see that by a judicious choice of the points x_i relative to $x = \pm 1/\sqrt{2}$ that H_{13} can also be made to vanish.

In our least squares example, if we make the transformation $x' = (x - 5)/5$ *and use a new model* $y = b_0' + b_1'x' + b_2'(2x'^2 - 1)$, *the normal equations become*

$$\begin{bmatrix} 10.0 & 0.0 & 0.0 \\ 0.0 & 5.0 & 0.0 \\ 0.0 & 0.0 & 5.0 \end{bmatrix} \begin{bmatrix} b_0' \\ b_1' \\ b_2' \end{bmatrix} = \begin{bmatrix} 15.33 \\ 4.62 \\ -0.21 \end{bmatrix}.$$

The diagonal matrix is trivial to invert and we get the result $b_0' = 1.533(59)$, $b_1' = 0.924(84)$, *and* $b_2' = -0.042(84)$.

The function we have chosen to be multiplied by the coefficient c is in fact the third member of a class of very useful functions known as *Chebychev polynomials*. They are defined by the general relation $y = T_m(x) = \cos\{n[\arccos(x)]\}$. In more practical terms, noting that $T_0(x) = 1$, and $T_1(x) = x$, they can be defined by the recursion relation $T_{m+1}(x) = 2xT_m(x) - T_{m-1}(x)$. Inspection of the definition shows that as x varies from -1 to $+1$, the polynomial oscillates between -1 and $+1$ n times. The Chebychev polynomials also have the interesting property that

$$\int_{-1}^{+1} \frac{T_m(x)T_n(x)}{(1-x^2)^{1/2}}\,dx = 0 \quad \text{if } m \neq n,$$

while

$$\int_{-1}^{+1} \frac{[T_n(x)]^2}{(1-x^2)^{1/2}}\,dx = \pi/2 \quad \text{if } n \geqslant 1.$$

Thus, if we wish to approximate a curve, $y = f(x)$, by a polynomial of the form

$$f(x) = \sum_{i=0}^{n} a_i T_i(x),$$

then

$$a_0 = (1/\pi)\int_{-1}^{+1}\left[f(x)/(1-x^2)^{1/2}\right]dx, \quad \text{and}$$

$$a_i = (2/\pi)\int_{-1}^{+1}\left\{[f(x)T_i(x)]/(1-x^2)^{1/2}\right\}dx \quad \text{for } i \geqslant 1.$$

Unlike a Taylor's series expansion, adding additional terms to improve the approximation does not affect the values of the coefficients of the preceding terms.

A third example of devising alternative parameters to avoid correlation problems is a case that arises frequently in crystallography when a crystal goes through a phase transition that doubles the size of the unit cell. Atoms that are related to each other by a vector that is one-half of the doubled cell translation are displaced by a small amount from their previous, more symmetric positions, and varying the position parameters of either will have exactly the same effect on the magnitudes of the structure factors. This will lead to catastrophic correlation effects if the position parameters are refined independently. The model may be made better conditioned if, instead of refining with x_1 and x_2 as parameters, we make $x' = x_1 + x_2$ and $x'' = x_1 - x_2$ the independent parameters, making use of the relationships

$$\frac{\partial F}{\partial x'} = \frac{\partial F}{\partial x_1} + \frac{\partial F}{\partial x_2} \quad \text{and} \quad \frac{\partial F}{\partial x''} = \frac{\partial F}{\partial x_1} - \frac{\partial F}{\partial x_2}.$$

The techniques we have described for avoiding correlation effects in model fitting all fall into the class of looking for linear combinations of parameters that are approximately eigenvectors of the Hessian matrix. It is usually not worthwhile looking for such changes of variable unless the correlation coefficients are very close to ± 1, i.e., $|\rho_{ij}| > 0.95$. When these conditions do arise, however, there are generally useful tricks to try, which include: (1) shift the origin; (2) use orthogonal polynomials instead of power series; and (3) refine sums and differences of correlated parameters.

Relationship between Precision and Accuracy

We have seen (page 97) that, if data points are weighted in proportion to the reciprocals of their variances, the inverse of the Hessian matrix of a function formed by the sum of the squares of the differences between observed values and those predicted by a model is the variance-covariance matrix for the adjustable parameters in the model. The diagonal elements of the variance-covariance matrix are the variances of the marginal distribution functions for those parameters. The square root of the estimated variance, the *estimated standard deviation*, or e.s.d., is a measure of the *minimum* possible width of a *confidence interval* within which the "correct" value of the associated parameter may safely be assumed to lie. The e.s.d. is a measure of *precision*. Of course, what the experimenter who made the measurements wants to know is, "How *accurate* is this measurement? What is the *maximum* value for the width of the interval such that I may safely assume that the 'correct' value of the parameter lies inside it?" We have also seen that if an independent measure of the precision of the data is available, statistical tests based on the F distribution may be used to judge whether the model gives an adequate description of the data. What can we assume about precision and accuracy if these tests force us to conclude that

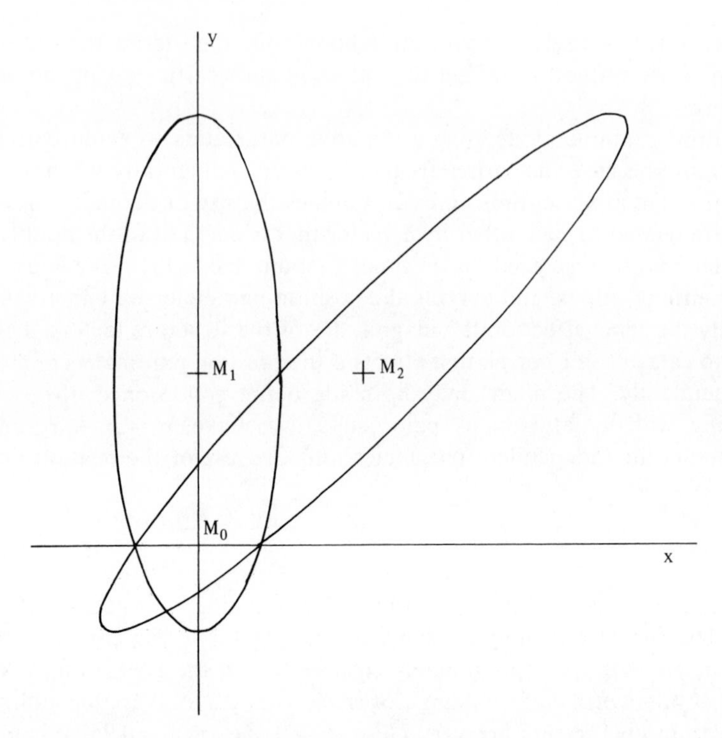

Fig. 8.2. Two possible lack-of-fit situations. The apparent minimum is at M_0. If the true minimum is at M_1, the estimate is unbiased. If it is at M_2, the estimate and its variance are both biased.

the model is not an adequate description of the data? The answer to this question depends critically on which of the two situations shown in Figure 8.2 we are dealing with.

The figure shows contours of the sum of squares as a function of one of the parameters we have refined, plotted horizontally, and the unknown parameter[2] that is missing from the model, plotted vertically. In one case the parameter we have estimated is statistically independent of the "missing" parameter, and the fact that we have not found the true minimum has no effect on the precision or accuracy of our measurement. In the other case, however, there is a strong correlation between the two parameters, and the failure to refine the missing one has not only caused us to underestimate the variance of the marginal distribution of our refined parameter, but it has also introduced a large *bias*, and the mean of the apparent distribution does not coincide at all with the mean of the true distribution.

[2]It makes no difference that the missing parameter may actually represent several physical effects. The parameter may be thought of, conceptually, as a linear function of several others whose direction in parameter space leads to the minimum.

One commonly used procedure for dealing with the problem of a sum of squares larger than that predicted is to assume that it is not due to lack of fit at all, but rather to a systematic underestimate of the variances of the individual measurements, leading to a corresponding overestimate of the appropriate weights. The quantity

$$\hat{\sigma}^2 = \left[\sum_{i=1}^{n} (R_i)^2 \right] \Big/ (n-p)$$

then is taken as an estimate of the true variance, and each element of the variance-covariance matrix is multiplied by it. A judgment of whether this procedure is appropriate may be made by making use of the fact that each term in the sum contributes, on the average, $[(n-p)/n]\sigma^2$ to the total. If we divide the total data set into subsets according to ranges of some experimental parameters, the quantity

$$\hat{\sigma}_k^2 = \left(\sum_{i=1}^{m_k} R_i^2 \right) \left[(n-p)/(nm_k) \right],$$

where m_k is the number of data points in subset k, should be an unbiased estimate of the overall variance, σ^2, and should therefore be the same for all subsets. If subsets are chosen in various different ways, and none of them shows a systematic variation as a function of the respective determining parameters, then we may be reasonably confident that the assumption of an overly optimistic estimate of the precision of the data is justified.

It often happens, however, that we have independent estimates of the precision of the data that cannot be reconciled with the assumption that the lack of fit is due to an underestimate of the data variances. In crystallography, for example, we may have measured the intensities of several sets of symmetry-equivalent reflections and found that the agreement among equivalent intensities is much better than the agreement between the data and the model. We cannot, then, justify blaming the lack of fit on the data, and we must assume that there is a deficiency in the model. A useful device in such cases is a scatter plot. Figure 8.3 shows a graph in which the vertical scale is the value of the standardized residual, $(y_i - \hat{y}_i)/\sigma_i$, and the horizontal scale is the value of some experimentally varied parameter. Each point in the data set is represented by a + in the appropriate spot on the graph. For a good model the +s should be randomly distributed on either side of the horizontal axis, and there should be no systematic trend as a function of the parameter being varied. In this case there is a clear trend from a predominance of positive values for the low values of x to a predominance of negative values for the high values of x, suggesting the presence of some unmodeled effect that is a function of this parameter. It is possible, then, to incorporate in the model a "fudge factor" that is a linear function of this parameter, and minimize the sum of squares again. An examination of the new row of the correlation matrix will reveal which of the other parameters

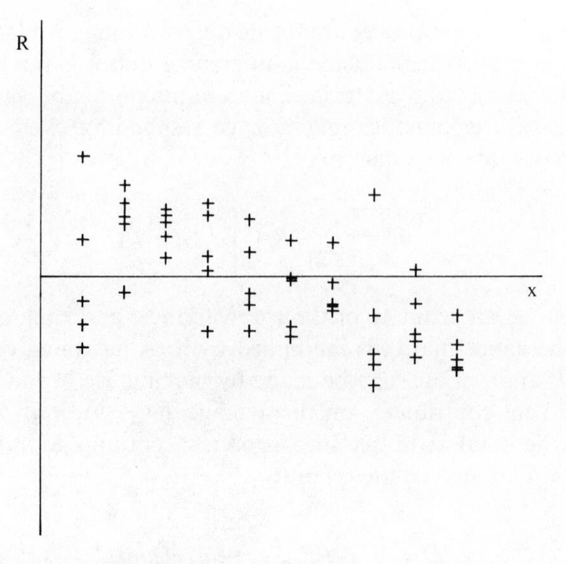

Fig. 8.3. A scatter plot of residuals as a function of a parameter, x, showing a systematic variation. Some effect depending on x may bias the parameter estimates.

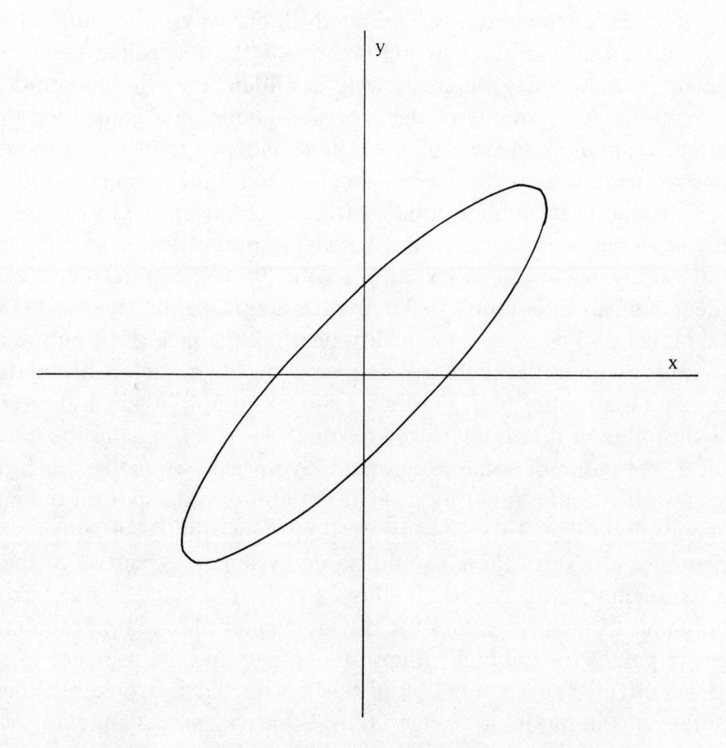

Fig. 8.4. Because of correlation between a parameter, x, and a hidden one, y, the variance of x may be underestimated, even though the fit is good.

are influenced by the new parameter. If correlations are small, we may assume (tentatively) that the precision is a reflection of corresponding accuracy, so that confidence limits may be set on the basis of statistical analysis. If there are significant correlations, however, it is necessary to examine carefully the physical nature of the new parameter. Is it a reflection of some undiscovered effect? Or is it a manifestation of some hitherto unnoticed systematic error?[3]

The reason that we have accepted our accuracy only tentatively is that we may now be in the situation shown in Figure 8.4. Again we show the contours of a discrepancy function as a function of a refined parameter and a hidden, unrefined one that is relatively poorly determined in our experiment, but for which we just happen to have assumed the correct value. There is no bias in the parameter estimate, but because we have neglected the correlation with the unseen parameter, we have greatly underestimated the variance of our refined parameter.

Note that in our least squares example, even though the coefficient of the quadratic term is essentially zero, it does make a difference to the estimates of the variances of the other two parameters whether or not the quadratic term is physically real. If we assume that the quadratic term must be identically zero, the inverse Hessian matrix becomes

$$\mathbf{H}^{-1} = \begin{pmatrix} 0.300 & -0.040 \\ -0.040 & 0.008 \end{pmatrix},$$

the estimated variance is $0.2551/8 = 0.03189$, *and the parameters are* $b_0 = 0.609(98)$, *and* $b_1 = 0.185(16)$. *These standard deviations are appreciably smaller than the corresponding ones given for the three-parameter fit on page* 103.

Uncertainties of Derived Functions: Propagation of Errors

When we have completed the process of fitting a model to a set of experimental data, and determined a variance-covariance matrix, $\mathbf{V}$, for our set of estimated parameters, $\hat{\mathbf{x}}$, we are usually interested in other quantities that are functions of the estimated parameters, and in the variances of these derived quantities. For example, we may have determined the position parameters of atoms in a crystal as fractions of the crystal lattice vectors, and we wish to derive a set of interatomic distances and bond angles, together with their respective variances.

[3] An observation, generally attributed to Enrico Fermi, is that nothing resembles a new phenomenon so much as a mistake.

Let us first consider derived quantities that are linear functions of the estimated parameters. We may express the derived quantity, y, by

$$\hat{y} = \sum_{j=1}^{p} a_j \hat{x}_j = \mathbf{a}^T \hat{\mathbf{x}}.$$

The variance of $\hat{y}$ is then given by

$$\langle (\hat{y} - y)^2 \rangle = \int \left[\mathbf{a}^T (\hat{\mathbf{x}} - \mathbf{x}) \right]^2 \Phi(\mathbf{x})\, d\mathbf{x},$$

where the integral is taken over all of parameter space, and $\Phi(\mathbf{x})$ is the joint probability density function for all parameters. This can be written

$$\langle (\hat{y} - y)^2 \rangle = \sum_{j=1}^{p} \sum_{k=1}^{p} a_j a_k \int (\hat{x}_j - x_j)(\hat{x}_k - x_k) \Phi(\mathbf{x})\, d\mathbf{x} = \mathbf{a}^T \mathbf{V} \mathbf{a}.$$

In general, if we have a set of derived quantities, represented by a vector, $\mathbf{y}$, related to the model parameters by $\mathbf{y} = \mathbf{A}\mathbf{x}$, then $\mathbf{y}$ has associated with it a variance-covariance matrix, $\mathbf{V}_y$, given by $\mathbf{V}_y = \mathbf{A}\mathbf{V}\mathbf{A}^T$.

Particular sets of derived quantities that are often of special interest are the set of predicted values for the data points, $\mathbf{y}_c$, and the set of residuals, $\mathbf{R}$, where $R_i = (y_{0i} - y_{ci})/\sigma_i$. In these cases the matrix $\mathbf{A}$ is the design matrix and, although the observed data can usually be correctly assumed to be uncorrelated, the calculated values—and therefore the residuals—are almost always correlated.

If y is not a linear function of $\mathbf{x}$, then the integral in

$$\langle (\hat{y} - y)^2 \rangle = \int \left[f(\mathbf{x}) \right]^2 \Phi(\mathbf{x})\, d\mathbf{x}$$

cannot, in general, be easily evaluated. We then can only approximate the variance of the derived quantity, which we do by expanding the function in a Taylor's series, dropping all but the linear terms, so that $a_i = \partial f(x)/\partial x_i$, and $\langle (\hat{y} - y)^2 \rangle \simeq \mathbf{a}^T \mathbf{V} \mathbf{a}$. The approximation will be good if the series expansion converges rapidly, and poor if it does not.

Chapter 9

Constrained Crystal Structure Refinement

In the previous chapters we presented a very general discussion of the mathematics of fitting a "model" to a set of data, but there has been very little discussion of the nature of the appropriate model, except for the implicit one that the first partial derivatives of the predicting function with respect to all parameters must exist. We shall now turn to the specific question of the determination of the parameters of a crystal structure by fitting diffraction data to a mathematical model of the structure.

The Observed Data

First it is necessary to discuss briefly what we mean by diffraction "data." In the diffraction of X-rays or neutrons by a single crystal, the actual observed quantity is the integrated area underneath a peak of scattered intensity due to Bragg reflection from a particular set of lattice planes. The peak rides on top of a smooth "background," and we are interested in the area between the peak and the background. The count of scattered intensity in each time interval is subject to a random variation according to a probability density function known as the *Poisson distribution*, which has the form $\Phi(n) = \exp(-\lambda)\lambda^n/n!$. Here $\Phi(n)$ is the probability of exactly n counts in a counting interval, and λ is a parameter that may be shown to be equal to the average number of counts per interval. The Poisson distribution is a discrete distribution function that has value zero for all negative values of n, a finite, positive value for $n = 0$, and positive values for positive values of n that increase to a maximum in the vicinity of $n = \lambda$, and then fall off increasingly rapidly as n becomes greater than λ. The distribution has a mean equal to λ and a variance also equal to λ, so that the standard deviation varies as $\lambda^{1/2}$, and the curve of n/λ becomes progressively sharper as λ becomes large. The variance of a sum of counts is equal to the sum of the counts, and the variance of the area between the peak and the background is the area under the peak, including the background, plus the variance of the background. The important point is that the observed intensity is subject to an unavoidable, random uncertainty due to the statistical variations in quantum counting. If the peak is measured by step scanning, with n_p steps over the peak with equal counting times, and the background is measured n_b times with the same counting time in each, the integrated intensity is $I = I_p - (n_p/n_b)I_b$, where I_p and I_b are the total counts in the peak and background regions, respectively. The variance of the integrated intensity is $\sigma_I^2 = I_p + (n_p/n_b)^2 I_b$.

The intensities of many reflections can be used directly as the observed data for a fit to the model but, in practice, this is rarely done. Instead, the observed number is taken to be "F squared," which is related to I by $F^2 = I/(kLP)$, or "F," which is the square root of F^2. Here L is the *Lorentz factor*, which is a function of the experimental geometry, P is the *polarization factor*, and k is a scale factor, which is sometimes measured experimentally, but is more often included as a variable parameter in the model. For unmonochromated X-rays $P = 1 + \cos^2 2\theta$ and is thus a function only of the scattering angle, θ, whereas for neutrons there are no polarization effects, so that $P = 1$. L and P are assumed to be measured with much greater precision than the statistical uncertainty in I, so that their uncertainty makes a negligible contribution to the variance of F^2, which is therefore given by $\sigma_{F^2}^2 = I/(kLP)^2$, while the variance of F is $\sigma_F^2 = \sigma_{F^2}^2/4F^2$.

Whether F^2 or F is used as the data value is largely a matter of local fashion. Most existing computer programs will handle either as run time options and, provided the proper expressions for the variances are used, the algorithms can be expected to give substantially identical results. Both numbers are functions only of experimentally measured quantities, and they are therefore statistically uncorrelated observations. Because the expressions are somewhat simpler, we shall assume in most of what follows that we are "refining on F;" that F is the number used as the "data."

The Model

The fundamental principle of physical optics is that, when a plane wave is incident on scattering matter, each point is the source of spherical waves that have a phase, relative to the phase of the plane wave at some arbitrary origin, equal to $(\boldsymbol{\kappa} \cdot \mathbf{r} + \pi)$, where $\mathbf{r}$ is the vector from the origin to the scattering point and $\boldsymbol{\kappa}$ is the *propagation vector* of the plane wave, a vector perpendicular to the plane wave front with a magnitude equal to the reciprocal of the wavelength. The spherical waves interfere constructively in certain directions to produce a new, "scattered," plane wave. The contribution of each point to the amplitude of the scattered wave is proportional to $\rho \exp[i(\boldsymbol{\kappa}_f - \boldsymbol{\kappa}_i) \cdot \mathbf{r})$, where ρ is a scattering density and $\boldsymbol{\kappa}_i$ and $\boldsymbol{\kappa}_f$ are the propagation vectors of the incident and scattered plane waves. If we designate by $\mathbf{h}$ the vector $(\boldsymbol{\kappa}_f - \boldsymbol{\kappa}_i)/2\pi$, we can write the amplitude of the scattered wave due to the scattering from a unit cell of a crystal in the form

$$F(\mathbf{h}) = \int \rho(\mathbf{r}) \exp(2\pi i \mathbf{h} \cdot \mathbf{r})\, d\mathbf{r},$$

where the integration is performed over the volume of the unit cell. The function $F(\mathbf{h})$, commonly known as the *structure factor*, although *structure amplitude* is probably a better name, is thus the Fourier transform of the scattering matter in the unit cell. For a crystal this function must be multiplied by the transform of a general, triply periodic function which, in a limit appropriate to a crystal containing a large number of unit cells, is equal to zero everywhere except at the nodes of the reciprocal lattice, where it may be assumed to be equal to one.

If we assume that the crystal is composed of atoms, recall that the Fourier transform of a Gaussian density function is also a Gaussian function, and also that the probability distribution of a particle in a harmonic potential well is Gaussian, we can write the structure factor as the sum of the transforms of the atoms, so it has the form

$$F(\mathbf{h}) = \sum_{j=1}^{n} f_j \exp\left(2\pi i \mathbf{h}^T \mathbf{r}_j - \mathbf{h}^T \beta_j \mathbf{h}\right)$$

where $\boldsymbol{\beta}_j$ is the temperature factor tensor for the jth atom, and the sum is taken over all of the atoms in the unit cell. Here $\mathbf{h} = h\mathbf{a}^* + k\mathbf{b}^* + l\mathbf{c}^*$ is given in reciprocal lattice units, and $\mathbf{r} = x\mathbf{a} + y\mathbf{b} + z\mathbf{c}$ is expressed as a vector whose components are fractions of the three lattice translations. f_j is the atomic scattering factor for the jth atom. For X-rays it is expressed as a multiple of the scattering power of a "Thomson electron," whereas for neutrons it is conventionally expressed as a length, in multiples of 10^{-14} meters. As a consequence, the density $\rho(\mathbf{r})$ has the units, for X-rays, of electrons per Å^3, and for neutrons it is a density of hypothetical nuclei with a scattering power of 1.0×10^{-14} meters. Because some nuclei, protons in particular, scatter with a phase shift of 0 rather than π, this density can be negative.

This expression for $F(\mathbf{h})$, when it is modified slightly by the inclusion of a scale factor and a function to describe the effects of extinction, may be termed the "conventional model." It is used, without restrictions except for those imposed by space-group symmetry, in the vast majority of cases of crystal structure refinement. The model in this form is, however, simultaneously too flexible and too restrictive.

The model is too flexible because it assumes implicitly that the adjustable parameters are independent of one another, whereas, in nature, there are relationships among them that are known to be invariant within a range that is much narrower than can be measured by any set of real data. Thus, for example, the position parameters of adjacent atoms are restricted by the fact that interatomic distances vary rather little; covalent bonds have well-defined lengths, and ionic radii are subject to variation over a rather narrow range. Further, the distances tend to be inflexible, so that the thermal motions of neighboring atoms are coupled in varying degrees, ranging up to the case in which groups of atoms vibrate as rigid bodies.

The model is too restrictive in that it assumes that the crystal is composed of discrete atoms, each of which lies in a potential well that is *harmonic*—i.e., that its potential energy is quadratic in displacements from an equilibrium position—in a rectilinear coordinate system. Rigid body motion, however, results in curvilinear motions for atoms near the periphery of the rigid body, and anharmonic forces can result in density distributions that are not Gaussian. Further, atomic disorder, caused, for example, by the incorporation in the structure of a molecule with less symmetry than the space-group site it occupies, also results in non-Gaussian density distributions. This *static disorder* is indistinguishable, in a diffraction experiment, from anharmonic motion.

We have seen, in our earlier discussion of moments and cumulants as examples of tensors, that the structure factor, in these more complicated cases, can be described by a more general formula involving third and/or fourth cumulants, but what we have said already about the excessive flexibility of the conventional model applies even more strongly to higher cumulant models: There are far too many mathematically independent parameters. Some wise elder statesman of experimental science observed, "With enough parameters you can fit an elephant!" G. S. Pawley drew on this aphorism and designated any parameter that did not correspond to a degree of freedom (in the physical, not statistical, sense) of the system an *elephant parameter*.

We shall proceed to discuss a number of techniques that may be used to apply *constraints* to the conventional model and its higher cumulant generalization in order to remove the elephant parameters and construct models that reflect what is known (by humans, but not by computers) about physics and chemistry. We should remark at the outset that, although the models we construct will be constrained relative to the generalized conventional model, the methods that are used to fit the model to the data are identical to those used for unconstrained refinement once the model has been formulated. This must be emphasized because the mention of the word "constraint" to a mathematician immediately calls to his mind certain techniques, such as the use of *Lagrange undetermined multipliers*, that are mathematically elegant but, like the expansion of determinants to invert matrices, computationally inefficient. Many an unsuspecting programmer has been led astray by consulting with such a mathematician.

The General Form for a Constrained Model

Let us represent the conventional model by $F(\mathbf{h}, \mathbf{x})$, where $\mathbf{x}$ represents a vector of p conventional parameters, including a scale factor, an extinction parameter, possibly occupancy factors that multiply the atomic scattering factor, position parameters, temperature factors, and maybe higher cumulants. The linearized Hessian matrix for the model is $\mathbf{H} = \mathbf{A}^T\mathbf{A}$, where

$$A_{ij} = \frac{\partial}{\partial x_j} F(\mathbf{h}_i, \mathbf{x}).$$

Now suppose the values of x_j are completely determined by some smaller number, q, of parameters, z_k, components of a vector $\mathbf{z}$. We clearly want a $q \times q$ Hessian matrix, $\mathbf{H}' = \mathbf{B}^T\mathbf{B}$, where $B_{ik} = (\partial/\partial z_k)F(\mathbf{h}_i, \mathbf{x})$. Remembering that if $f(y)$ is a function of y and y is a function of x, that $(\partial f/\partial x) = (\partial f/\partial y)(\partial y/\partial x)$, we can write

$$\frac{\partial F(\mathbf{h}_i, \mathbf{x})}{\partial z_k} = \sum_{j=1}^{p} \frac{\partial F(\mathbf{h}_i, \mathbf{x})}{\partial x_j} \frac{\partial x_j}{\partial z_k},$$

or, in matrix form, $\mathbf{B} = \mathbf{AC}$, where $C_{jk} = \partial x_j/\partial z_k$ forms a matrix with p rows and q columns. $\mathbf{C}$ is called the *constraint matrix*.

The least squares solution (after iterating to convergence) is, as before, $\hat{\mathbf{z}} = (\mathbf{B}^T\mathbf{WB})^{-1}\mathbf{B}^T\mathbf{WR} = (\mathbf{C}^T\mathbf{A}^T\mathbf{WAC})^{-1}\mathbf{C}^T\mathbf{A}^T\mathbf{WR}$, where $\mathbf{W}$ is the weight matrix, and $\mathbf{R}$ is the vector of residuals, $(F_0 - F_c)$. The variance-covariance matrix of conventional parameters is $\mathbf{V} = \mathbf{C}(\mathbf{B}^T\mathbf{WB})^{-1}\mathbf{C}^T$, and may be used to develop a variance-covariance matrix for any derived quantities by procedures we have already discussed.

The question of whether the conventional model gives a "significantly" better fit to the data than a constrained model may be addressed by refining both models. Designating the weighted sums of squares of residuals for the constrained and unconstrained models by S_c^2 and S_μ^2, respectively, the F ratio,

$$F = \left[(S_c^2 - S_u^2)/(p - q)\right] / \left[S_u^2/(n - p)\right],$$

may be compared with the cumulative F distribution with $(p - q)$ and $(n - p)$ degrees of freedom to find the probability that the ratio would have that value or greater by chance alone. A *necessary* condition for significance is that this probability be small. Sufficiency, as we have seen previously, depends on the appropriateness of the model.

An alternative test for significance that has a certain popularity among crystallographers was developed by W. C. Hamilton and is therefore known as *Hamilton's R-factor ratio test*. It makes use of the *weighted R index*, defined by

$$R_w = \left[\sum_{i=1}^{n} w_i (F_{0i} - F_{ci})^2 \Big/ \sum_{i=1}^{n} w_i F_{0i}^2 \right]^{1/2}.$$

Clearly $R_w^2 = S^2/\sum_{i=1}^{n} w_i F_{0i}^2$, and therefore $[(R_c^2 - R_u^2)/(p - q)]/[R_u/(n - p)]$ also has the F distribution with $(p - q)$ and $(n - p)$ degrees of freedom. A little bit of algebraic manipulation shows that the ratio R_c/R_u has the distribution function

$$\Phi(R_c/R_u) = \{[(p-q)/(n-p)]\Phi_F[(p-q),(n-p)]+1\}^{1/2},$$

which may be readily computed from the F distribution.

Shape Constraints

One of several types of constraint that are appropriate in crystallography is a *shape constraint*. Many molecules have an inherent symmetry that is not reflected in the point-group symmetry of the crystallographic sites they occupy. Also, many structures consist of several units whose structure is known, as in the construction of polypeptides from amino acid subunits. A particularly important case is that of a planar molecule occupying a site containing no mirror plane. In each of these cases it is convenient to describe the molecule in terms of a coordinate system that is fixed with respect to the molecule and that reflects the molecular symmetry. Then, in addition to the coordinates of the atoms in this coordinate system, three parameters are needed to locate the origin of the coordinate system, and three more to specify its orientation with respect to the crystal axes. Thus, for example, we might describe a tetrahedral SiO_4 group by the three coordinates, in the crystal system, of the silicon atom, three Eulerian angles relating the twofold axes of the group to the crystal axes, and the Si—O bond length, a total of 7 parameters rather than the 15 that are needed if all atoms are in general positions.

For another example, a planar, five-membered ring can be described in a system in which the origin lies on one atom, and another atom defines the x axis, while all five atoms are at $z = 0$. In this case we have three coordinates for the origin, three angles to specify the orientation, one parameter for the atom on the x axis, and two parameters each for the other three atoms, for a total of 13 parameters, as compared with 15 for the unconstrained model. If there is additional symmetry, such as a mirror plane perpendicular to the plane of the molecule, the parameters are restricted still further, to a total of only 10 for the five-membered ring.

In implementing such a system of constraints it is convenient to make use of a standard, orthonormal coordinate system that is fixed with respect to the crystal axes (see Fig. 9.1), with its x axis parallel to $\mathbf{a}$, its y axis lying in the $\mathbf{a}-\mathbf{b}$ plane, and its z axis parallel to $\mathbf{a}\times\mathbf{b}$. The linear transformation that converts a vector in this coordinate system to the crystal axis system is $\mathbf{B}$, the upper triangular matrix such that $\mathbf{B}\mathbf{B}^T = \mathbf{G}^{-1}$, where $\mathbf{G}^{-1}$ is the reciprocal lattice metric tensor. The Eulerian angle ω is a rotation about the z axis of the standard system required to bring the z axis of the special system into the $x-z$ plane of the standard system. A vector in a coordinate system rotated by ω with respect to the standard system is transformed

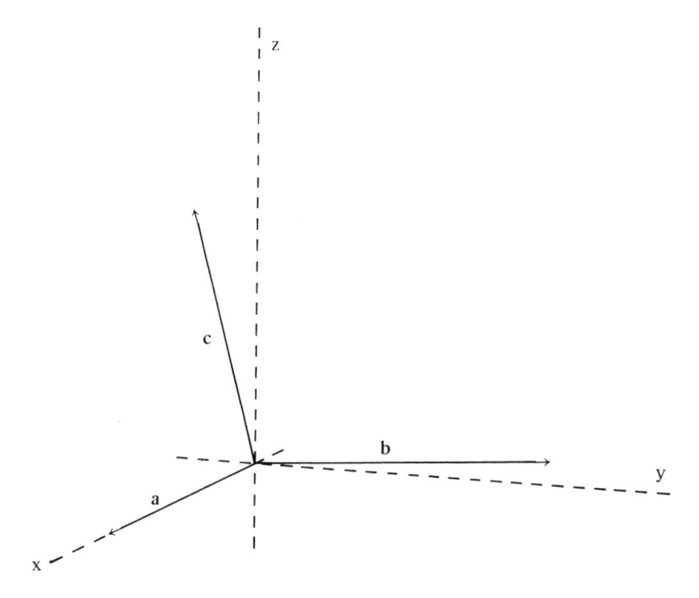

Fig. 9.1. The relationship between the crystal axes, **a**, **b**, and **c**, and the standard, orthonormal coordinate system.

to the standard system by the matrix

$$\mathbf{\Omega} = \begin{pmatrix} \cos\omega & -\sin\omega & 0 \\ \sin\omega & \cos\omega & 0 \\ 0 & 0 & 1 \end{pmatrix}.$$

The angle χ is a clockwise rotation about the y axis of the standard system (viewed down the positive y axis) to bring the z axes of the two coordinate systems into coincidence. A vector in the coordinate system rotated by χ is transformed into the ω system by the matrix

$$\mathbf{X} = \begin{pmatrix} \cos\chi & 0 & \sin\chi \\ 0 & 1 & 0 \\ -\sin\chi & 0 & \cos\chi \end{pmatrix}.$$

The angle φ is a clockwise rotation about the common z axis to bring the x and y axes into coincidence. A vector in the coordinate system rotated by φ is transformed to the χ system by the matrix

$$\mathbf{\Phi} = \begin{pmatrix} \cos\varphi & -\sin\varphi & 0 \\ \sin\varphi & \cos\varphi & 0 \\ 0 & 0 & 1 \end{pmatrix}.$$

A vector in the special coordinate system is therefore transformed into the standard, orthonormal system by the matrix $\mathbf{R} = \mathbf{\Omega X \Phi}$, and into the crystal, fractional system by the matrix $\mathbf{T} = \mathbf{BR}$. A FORTRAN subroutine given in Appendix G generates the matrix $\mathbf{R}$ together with its derivatives with respect to the Eulerian angles. These relationships may be used to generate the derivatives of F_c with respect to atom coordinates referred to the special system and with respect to the orientation angles of the special system.

Rigid-Body Thermal Motion Constraints

We saw in Chapter 5 that the displacement, $\mathbf{u}$, of a point in a rigid body due to a rotation of the rigid body described by the axial vector $\mathbf{\Lambda}$ can be written exactly in the form

$$\mathbf{u} = (\sin\lambda/\lambda)(\mathbf{\Lambda} \times \mathbf{r}) + \left[(1 - \cos\lambda/\lambda^2)\right]\left[\mathbf{\Lambda} \times (\mathbf{\Lambda} \times \mathbf{r})\right]$$

where λ represents $|\mathbf{\Lambda}|$ and $\mathbf{r}$ is a vector from a fixed point, the center of rotation, in the rigid body to the point under consideration. If the center of rotation is displaced from its equilibrium position by the vector $\mathbf{t}$, the total displacement of the point $\mathbf{r}$ is $\mathbf{v} = \mathbf{u} + \mathbf{t}$. The cumulants of the distribution of $\mathbf{v}$, which appear in the generalized structure factor formula, can be written in terms of moments of the same and lower orders. Thus, if we can express the moments in terms of general parameters, such as the elements of the rigid-body motion tensors, $\mathbf{T}$, $\mathbf{L}$, and $\mathbf{S}$, the tensor elements can be the independent parameters determined in a refinement.

We shall assume that the angle, λ, is small enough for $\sin\lambda$ and $\cos\lambda$ to be replaced by their power series expansions, truncated after the terms involving λ^4. We shall also make the approximation that the joint probability density function of λ_1, λ_2, λ_3, t_1, t_2, and t_3 is Gaussian. As we have seen, this is a sufficient, although not necessary, condition for all odd moments—terms of the form $\langle\lambda_i\lambda_j\lambda_k\rangle$, $\langle\lambda_i t_j t_k\rangle$, etc.—to vanish. It also enables us to approximate fourth moments in terms of the second moments, that is, in terms of the rigid-body motion tensor elements.

The first moment is a vector whose components are

$${}^1\mu_i = \langle v_i\rangle = \langle t_i\rangle + \sum_{j=1}^{3}\left\{A_{ij}\langle\lambda_j\rangle + \sum_{k=1}^{3}\left[B_{ijk}\langle\lambda_j\lambda_k\rangle + \sum_{l=1}^{3}\left(C_{ijkl}\langle\lambda_j\lambda_k\lambda_l\rangle + \sum_{m=1}^{3} D_{ijklm}\langle\lambda_j\lambda_k\lambda_l\lambda_m\rangle\right)\right]\right\},$$

where A, B, C, and D were defined in Chapter 5. Using our approximations, this reduces to

$${}^1\mu_i = \sum_{j=1}^{3}\sum_{k=1}^{3}\left[B_{ijk}L_{jk} + \sum_{l=1}^{3}\sum_{m=1}^{3} D_{ijklm}\left(L_{jk}L_{lm} + L_{jl}L_{km} + L_{jm}L_{kl}\right)\right].$$

This very important vector gives the displacement from the equilibrium position of an atom undergoing curvilinear motion to the position of the mean of its time-averaged distribution function. Its effect causes an apparent shrinking of a rigid molecule as the temperature, and therefore the average libration amplitude, is increased.

Because the coefficients B_{ijk} and D_{ijklm} are all multiples of the equilibrium position vector, **r**, the first moment can be written ${}^{1}\boldsymbol{\mu} = \mathbf{Mr}$, where **M** is a matrix whose elements are functions of the elements of **L**. The apparent position of the atom, **x**, is $\mathbf{r} + {}^{1}\boldsymbol{\mu}$, so that $\mathbf{x} = (\mathbf{I} + \mathbf{M})\mathbf{r}$. The equilibrium position, which is what we really want to know, is therefore given by $\mathbf{r} = (\mathbf{I} + \mathbf{M})^{-1}\mathbf{x}$. This linear transformation is known as the *libration correction.*

The higher moments, and therefore the higher cumulants, of the atomic distribution can be determined by evaluating expressions of the type $\langle v_i v_j \rangle$, $\langle v_i v_j v_k \rangle$, etc., giving the moments and cumulants as functions of the elements of **T**, **L**, and **S**. The algebra is straightforward but tedious, and it would serve no useful purpose to write it all out. FORTRAN expressions for second and third cumulants, as well as their derivatives with respect to the elements of **T**, **L**, and **S**, are given in Appendix G.

If a further approximation is made, and all of the expressions are truncated after the terms that are quadratic in t_j and λ_j, indeterminacies enter under certain conditions. The diagonal elements of **S**—terms of the form $\langle \lambda_i t_i \rangle$—do not appear by themselves, but only as differences of pairs. Therefore a constant added to all three will leave the cumulants calculated from them unchanged. This is the so-called *trace-of-***S** *singularity*. It is usually dealt with by setting $S_{11} + S_{22} + S_{33} = 0$. Further, if a planar group has all of its members lying on a curve that is one of the conic sections, circle, ellipse, parabola, or hyperbola (and this is always possible if the group has five or fewer members), the transformation connecting their individual cumulants with the elements of **T** and **L** is singular, and the relations are not unique. Although neither the trace-of-**S** problem nor the conic section problem arises strictly when we use the expressions we have derived, since they contain fourth-degree terms that do not vanish, the matrices involved do, nevertheless, tend to be very nearly singular, and the procedure is ill-conditioned. It is therefore good practice to assume that the limitations of the **TLS** analysis still apply even with the better approximations provided by these formulas.

This is, perhaps, a good place to discuss two practices that are frequently found in the analysis of refined structures: the fitting of rigid-body parameters to the refined values of anisotropic temperature factors, and the fitting of a "least squares mean plane" to a nearly planar group that has not been constrained to be planar, coupled with an examination of the distances of various atoms from this plane in order to judge whether the group is "significantly" nonplanar. Both of these procedures are designed to answer the wrong question. The questions that should be asked, and they can only be addressed through properly designed constrained refinements, are,

"What is the best rigid-body fit to the data?" and, "Do the data indicate that the group is or is not planar?" In both cases there are other parameters that must be allowed to find their best values consistent with the constrained model, and it is only this adjusted model that can legitimately be compared with the unconstrained one.

Chemical Constraints

A third type of constraint is very useful in cases where a crystal structure contains chemical disorder. One common situation arises when a crystal contains a molecular group that is almost symmetrical, except that one orientation has a nitrogen atom in the position that would be a carbon atom for the other orientation, and vice versa. Another common situation occurs in minerals and inorganic compounds that contain mixtures of cations, frequently transition metals, that can readily substitute for one another on various nonequivalent crystallographic sites. In both cases the overall chemical composition is known, and any refinement of the occupancies of the respective sites must reflect this chemical knowledge.

Suppose we have N_a species of atom distributed among N_s distinct sites with multiplicities, in a unit cell, of m_i, where $i = 1, 2, \ldots, N_s$. If we designate by A_{ij} the fraction of sites of type i that is occupied by atomic species j, we can write a system of N_a equations of the form

$$\sum_{i=1}^{N_s} A_{ij} m_i = C_j,$$

where C_j is the total concentration of atoms of species j in the unit cell, as determined by chemical analysis. We wish to determine the individual fractions, A_{ij}, of each atomic species at each site, but they are not all independent. We can solve each of these equations for one of its coefficients, such as that for $i = 1$,

$$A_{ij} = (1/m_1)\left(C_j - \sum_{i=2}^{N_s} A_{ij} m_i \right),$$

and we can then modify the partial derivative of F_c with respect to A_{ij}, where $i = 2, 3, \ldots, N_s$, by adding a term $(\partial F_c / \partial A_{ij})(\partial A_{ij} / \partial A_{1j}) = -(\partial F_c / \partial A_{ij})(m_i / m_1)$. Because each atom that is removed from one site must be added to another, this procedure greatly reduces the correlation that is otherwise observed between occupancy factors and temperature factors, and thereby enhances the precision with which occupancies may be determined.

Representing non-Gaussian Distributions

We have seen that the characteristic function, or Fourier transform, of the Gaussian density function, $G(x, \mu, \sigma) = (2\pi\sigma^2)^{-1/2}\exp[-(1/2)\cdot(x-\mu)^2/\sigma^2]$, has the form $\Psi(t) = \exp[i\,{}^1\kappa t - (1/2)^2\kappa t]$, where ${}^1\kappa$ and ${}^2\kappa$ are the first and second cumulants of the distribution. We have also suggested that distributions that are almost Gaussian can be described by adding one or two more terms in the argument of the exponential in this characteristic function, thereby treating the argument as a power series in it. We shall now discuss, briefly, how we may determine what the shape of a probability density function with nonvanishing third and/or fourth cumulants actually looks like.

Let us first consider the standardized Gaussian probability density function, $\gamma(x) = (2\pi)^{-1/2}\exp[-(1/2)x^2]$, and its derivatives with respect to x.

$$\frac{d\gamma}{dx} = -(2\pi)^{-1/2}x\exp\left[-(1/2)x^2\right] = -x\gamma(x).$$

$$\frac{d^2\gamma}{dx^2} = (2\pi)^{-1/2}(x^2-1)\exp\left[-(1/2)x^2\right] = (x^2-1)\gamma(x).$$

It is apparent that the nth derivative will be a polynomial of degree n in x multiplying $\gamma(x)$. These polynomials are a class known as *Hermite polynomials*. They are commonly designated $H_n(x)$. The general form for $H_n(x)$ is

$$H_n(x)\gamma(x) = (-1)^n\frac{d^n}{dx^n}\gamma(x).$$

The first few Hermite polynomials are

$$H_0(x) = 1,$$

$$H_1(x) = x,$$

$$H_2(x) = x^2 - 1,$$

$$H_3(x) = x^3 - 3x,$$

$$H_4(x) = x^4 - 6x^2 + 3.$$

(Warning! There are *two* sets of related but distinct polynomials, both of which are called Hermite polynomials. Those given here are used by statisticians because of their close relationship to the Gaussian density function. The other set appears in quantum mechanics in the solutions to Schrödinger's equation for a harmonic potential function. They are discussed in Appendix E.) To derive some useful properties of these polynomials, consider the function $\gamma(x-t)=(2\pi)^{-1/2}\exp[-(1/2)(x^2-2xt+t^2)]$ $=\gamma(x)\exp(tx-t^2/2)$. We can expand this function in a Taylor's series, giving

$$\gamma(x-t)=\sum_{j=0}^{+\infty}\left[(-1)^j/j!\right]t^j\frac{d^j}{dx^j}\gamma(x)$$

$$=\sum_{j=0}^{+\infty}(t^j/j!)H_j(x)\gamma(x).$$

Because $\gamma(x)$ is positive for all x we then get

$$\exp(tx-t^2/2)=\sum_{j=0}^{+\infty}(t^j/j!)H_j(x).$$

Differentiating both sides of this equation with respect to x, we get

$$t\exp(tx-t^2/2)=\sum_{j=0}^{+\infty}(t^j/j!)\frac{d}{dx}H_j(x),$$

or

$$\sum_{j=0}^{+\infty}(t^{j+1}/j!)H_j(x)=\sum_{j=0}^{+\infty}(t^j/j!)\frac{d}{dx}H_j(x).$$

For this equation to be satisfied for all values of t, the coefficients of each power of t must be individually identical, giving

$$H_{r-1}(x)/(r-1)!=\frac{d}{dx}H_r(x)/r!$$

or

$$\frac{d}{dx}H_r(x)=rH_{r-1}(x).$$

From the general, defining formula

$$\int_{-\infty}^{+\infty}H_n(x)\gamma(x)\,dx=-\left[H_{n-1}(x)\gamma(x)\right]_{-\infty}^{+\infty}=0$$

if $n\geqslant 1$. If $n=0$,

$$\int_{-\infty}^{+\infty}H_0(x)\gamma(x)\,dx=\int_{-\infty}^{+\infty}\gamma(x)\,dx=1.$$

Using these relationships we consider integrals of the form

$$\int_{-\infty}^{+\infty}H_m(x)H_n(x)\gamma(x)\,dx,\quad\text{where } m\leqslant n.$$

Setting $u = H_m(x)$ and $dv = H_n(x)\gamma(x)\,dx$, we integrate by parts, getting

$$\int_{-\infty}^{+\infty} H_m(x)H_n(x)\gamma(x)\,dx = -\left[H_m(x)H_{n-1}(x)\gamma(x)\right]_{-\infty}^{+\infty} + m\int_{-\infty}^{+\infty} H_{m-1}(x)H_{n-1}(x)\gamma(x)\,dx.$$

The expression in the square brackets vanishes at both limits, and thus is equal to zero. The remaining integral is identical to the one we started with, but with the degrees of both polynomials reduced by one. If we repeat the integration by parts $(m-1)$ more times, we will get to

$$m!\int_{-\infty}^{+\infty} H_0(x)H_{n-m}(x)\gamma(x)\,dx.$$

We saw above that this integral vanishes unless $n = m$, in which case it is equal to one. Therefore

$$\int_{-\infty}^{+\infty} H_m(x)H_n(x)\gamma(x)\,dx = 0 \quad \text{if } m \neq n.$$

$$\int_{-\infty}^{+\infty} \left[H_n(x)\right]^2\gamma(x)\,dx = n!.$$

Let us now consider a probability density function, $\Phi(x)$, and let us assume that it can be approximated by an expression of the form

$$\Phi(x) = \gamma(x)\left[\sum_{i=0}^{+\infty} a_i H_i(x)\right].$$

If $a_0 = 1$, and $a_i = 0$ for all $i > 0$, $\Phi(x)$ is a Gaussian density function, and we might hope that, if $\Phi(x)$ is similar to a Gaussian function, that the series will converge rapidly, and that only a few terms will be needed. We can determine the appropriate value of the jth coefficient, a_j, by multiplying both sides of the equation by $H_j(x)$, and integrating, giving

$$\int_{-\infty}^{+\infty} \Phi(x)H_j(x)\,dx = \int_{-\infty}^{+\infty} H_j(x)\left[\sum_{i=0}^{+\infty} a_i H_i(x)\right]\gamma(x)\,dx,$$

but all terms in the sum on the right-hand side vanish except for the one for which $i = j$, which is equal to a $j!$. Therefore

$$a_j = (1/j!)\int_{-\infty}^{+\infty} H_j(x)\Phi(x)\,dx.$$

Now $H_j(x)$ is a polynomial of degree j in x, and terms of the form

$$\int_{-\infty}^{+\infty} x^n\Phi(x)\,dx$$

are the moments of the density function $\Phi(x)$. If we substitute the explicit expressions for the first few Hermite polynomials, we get

$$a_0 = 1,$$

$$a_1 = {}^1\mu,$$

$$a_2 = (1/2)({}^2\mu - 1)\left[= (1/2)(\sigma^2 - 1)\right],$$

$$a_3 = (1/6)({}^3\mu - 3{}^1\mu),$$

$$a_4 = (1/24)({}^4\mu - 6{}^2\mu + 3).$$

If we now make the substitution $x' = (x - \mu)/\sigma$, so that the variance of our distribution, $\Phi(x')$, is equal to 1 and its mean is equal to 0, we get

$$a_0 = 1,$$

$$a_1 = 0,$$

$$a_2 = 0,$$

$$a_3 = (1/6)^3\mu = (1/6)^3\kappa,$$

$$a_4 = (1/24)({}^4\mu - 3) = (1/24)^4\kappa,$$

and our approximation for $\Phi(x)$ becomes

$$\Phi(x) = \gamma(x')\left\{1 + ({}^3\mu/6)H_3(x') + \left[({}^4\mu - 3)/24\right]H_4(x') + \cdots\right\},$$

or

$$\Phi(x) = \gamma(x')\left[1 + ({}^3\kappa/6)H_3(x') + ({}^4\kappa/24)H_4(x') + \cdots\right].$$

These expressions are identical. The one expressed in terms of moments is known as a *Gram-Charlier series*. The one expressed in terms of cumulants is known as an *Edgeworth series*. If the series includes moments or cumulants up to the fourth, it gives a function that has the same values for the integrals that define the first four moments as the density function $\Phi(x)$. It can be expected, in practice, to give a reasonably good approximation to $\Phi(x)$ in the peak region, where the function has most of its area. However, because the $H_n(x)$ functions are polynomials, even if the coefficients are small, the highest degree term will always become dominant at a large value of x, and so the tails will not necessarily fit well. In particular, the function will always be asymptotic to the x axis at large values of x with the sign of the highest degree term. Thus, if the last term retained is cubic, the function will be negative either at large positive x or large negative x, and it cannot be an acceptable probability density function. If the last term retained is biquadratic, the function can be everywhere positive, but only if the fourth cumulant is positive (positive kurtosis, meaning that the tails contain more area than a Gaussian). If the fourth cumulant is negative, both tails will be negative.

An important application of the Edgeworth series arises when we wish to represent the actual shape of the distribution of scattering density due to an atom that is subject to anharmonic forces, or to static disorder. We assume that the proper statistical tests have been applied, and that the results have been confirmed by physical intuition and all other available evidence, so that we can be confident that third and/or fourth cumulants are "significantly" different from zero. We can generalize the Hermite polynomials to three dimensions by defining them in a manner similar to the one we used in one dimension.

$$H_{ijk}(x, y, z)G(x, y, z) = (-1)^{i+j+k}\frac{\partial^{i+j+k}}{\partial x^i \partial y^j \partial z^k}G(x, y, z),$$

where $G(x, y, z)$ is the trivariate Gaussian probability density function. If we assume that we are interested in the distribution of one atom at a time, we can express the distribution function in terms of modification of a Gaussian density function that has the same variance-covariance matrix as our unknown function. Further, we can choose the coordinate axes along the eigenvectors of the variance-covariance matrix and, again, scale the axes by the respective eigenvalues, so that the variance along each axis is equal to 1. The function $G(x, y, z)$ is then the product of three, univariate Gaussian functions.

$$G(x, y, z) = \gamma(x)\gamma(y)\gamma(z).$$

Each partial derivative is then the product of three single-variable Hermite polynomials—one a function of x only, one a function of y only, and one a function of z only. The three-dimensional Edgeworth series then becomes

$$\Phi(x, y, z) = \gamma(x)\gamma(y)\gamma(z)\Bigg\{1 + (1/6) \times \sum_{i=1}^{3}\sum_{j=1}^{3}\sum_{k=1}^{3}\Bigg[{}^3\boldsymbol{\kappa}_{ijk}H_m(x)H_n(y)H_p(z) + (1/4)\sum_{l=1}^{3}{}^4\boldsymbol{\kappa}_{ijkl} \times H_m(x)H_n(y)H_p(z) + \cdots\Bigg]\Bigg\},$$

where m, n, and p have values from 0 to 4 depending on the number of times 1, 2, and 3, respectively, appear among i, j, k, and l.

Appendix A: Eigenvalues and Eigenvectors of 3×3 Symmetric Matrices

The determination of eigenvalues and eigenvectors of symmetric matrices in three-dimensional space requires the solution of the secular equation $|\mathbf{A} - \lambda\mathbf{I}| = 0$. This is a cubic equation of the form $\lambda^3 + a\lambda^2 + b\lambda + c = 0$, where

$$a = -(A_{11} + A_{22} + A_{33}),$$

$$b = (A_{11}A_{22} - A_{12}^2) + (A_{11}A_{33} - A_{13}^2) + (A_{22}A_{33} - A_{23}^2), \quad \text{and}$$

$$c = -(A_{11}A_{22}A_{33} + 2A_{12}A_{13}A_{23} - A_{11}A_{23}^2 - A_{22}A_{13}^2 - A_{33}A_{12}^2).$$

Although it is rarely discussed in standard algebra courses, there is a direct algorithm for the analytic solution of cubic equations. The first step is to make the substitution $\lambda = x - a/3$, which eliminates the coefficient of the quadratic term and leaves an equation of the form $x^3 - qx - r = 0$, where $q = a^2/3 - b$, and $r = -(2a^3/27 - ab/3 + c)$. If the matrix is symmetric, the equation must have three real roots, meaning that $27r^2$ must be less than $4q^3$. We find the smallest positive angle $\varphi = \arccos[(27r^2/4q^3)^{1/2}]$, and the three eigenvalues are then $\lambda_1 = C\cos\varphi/3 - a/3$, $\lambda_2 = -C \cdot \cos[(\pi - \varphi)/3] - a/3$, and $\lambda_3 = -C\cos[(\pi + \varphi)/3] - a/3$, where $C = (4q/3)$.

When the three eigenvalues have been determined, the eigenvectors may be found by substituting the eigenvalues into the system of equations $\mathbf{Au} = \lambda\mathbf{u}$. However, because the matrix $(\mathbf{A} - \lambda\mathbf{I})$ is singular when λ is an eigenvalue, only two of the three equations are independent. The third condition is supplied by requiring that the coordinate system defined by the three eigenvectors be orthonormal. Let u_{1i}, u_{2i}, and u_{3i} be the direction cosines of the eigenvector corresponding to the eigenvalue λ_i, with respect to the coordinate system in which $\mathbf{A}$ is defined. We then have $u_{1i}^2 + u_{2i}^2 +$

$u_{3i}^2 = 1$. Let x_1 and x_2 be u_{1i}/u_{3i} and u_{2i}/u_{3i}, respectively. (This assumes that $u_{3i} \neq 0$. If two of the off-diagonal elements in any row and column of **A** are equal to zero, then one of the coordinate axes is an eigenvector, and the problem reduces to a quadratic.) The basic equations can then be written

$$(A_{11} - \lambda_i)x_1 + A_{12}x_2 = -A_{13},$$

$$A_{12}x_1 + (A_{22} - \lambda_i)x_2 = -A_{23}.$$

The solutions are

$$x_1 = \left[-A_{13}(A_{22} - \lambda_i) + A_{12}A_{23}\right]/\Delta,$$

$$x_2 = \left[-A_{23}(A_{11} - \lambda_i) + A_{12}A_{13}\right]/\Delta,$$

where $\Delta = [(A_{11} - \lambda_i)(A_{22} - \lambda_i) - A_{12}^2]$. The direction cosines are then

$$u_{1i} = x_1/D, \quad u_{2i} = x_2/D, \quad \text{and } u_{3i} = 1/D,$$

where $D = (x_1^2 + x_2^2 + 1)^{1/2}$.

If the elements of an anisotropic temperature factor tensor, $\boldsymbol{\beta}$, have been determined by least squares refinement, this procedure enables us to compute the directions and magnitudes of the principal axes of the thermal ellipsoid. We would also like to know something about the precision with which we have determined these amplitudes. We will have determined the variance-covariance matrix for the six independent tensor elements, β_{ij}. Denote this 6 × 6 matrix by **V**. We need the vector of derivatives, **D**, whose elements are $D_i = \partial\lambda_i/\partial\beta_{jk}$. Then the variance of λ, σ_λ^2, is given by $\sigma_\lambda^2 = \mathbf{D}^T\mathbf{V}\mathbf{D}$. The expressions for the derivatives, D_i, are complicated, but each step in the determination of λ expresses a set of derived quantities as an analytic function of previously derived quantities, so that the required derivatives can be readily represented by a matrix product in which the elements of each matrix are the derivatives of derived quantities with respect to the quantities derived in the previous step.

Appendix B: Stereographic Projection

Not too many years ago all students who were introduced to crystallography, even if they did not intend to study it in any great depth, were initiated with a few laboratory exercises in the use of stereographic projections. In recent years there seems to have been a tendency to neglect this topic, which is unfortunate because a knowledge of the techniques of drawing and interpreting stereographic projections is extremely useful.

The normals to planes in a crystal, whether they are natural faces or invisible reflecting planes, may be projected onto the surface of a sphere by locating the crystal at the center of the sphere, drawing radial lines perpendicular to the planes, and marking the points at which they intersect the surface. Referring to Figure A.1, the points marked on the surface of the sphere may be represented on a plane sheet of paper by drawing a line from the point to the opposite pole of the sphere, and marking the intersection of this line with the equatorial plane. All points on the upper hemisphere may thus be plotted inside the equatorial circle. We represent an actual crystal in projection by considering that the polar axis of our sphere is perpendicular to our paper. Meridians of longitude are then represented by radial lines in the circle.

Figure A.2 shows a section through the sphere containing the polar axis and a point on the surface with colatitude θ. It is apparent that all points on a parallel of latitude will appear in projection on a circle concentric with the equatorial circle, with a radius given by $r = R\tan(\theta/2)$, where R is the radius of the equatorial circle. It is obvious that all points in the hemisphere above the plane of the paper will project as points within the equatorial circle. Points on the lower hemisphere would project outside the circle, and it is occasionally useful to represent them that way (such as when they lie on a small circle about a point near the equator), but they are usually represented by projecting them through the upper pole of the sphere and marking them by small open rings instead of by dots. Figure A.3 shows a

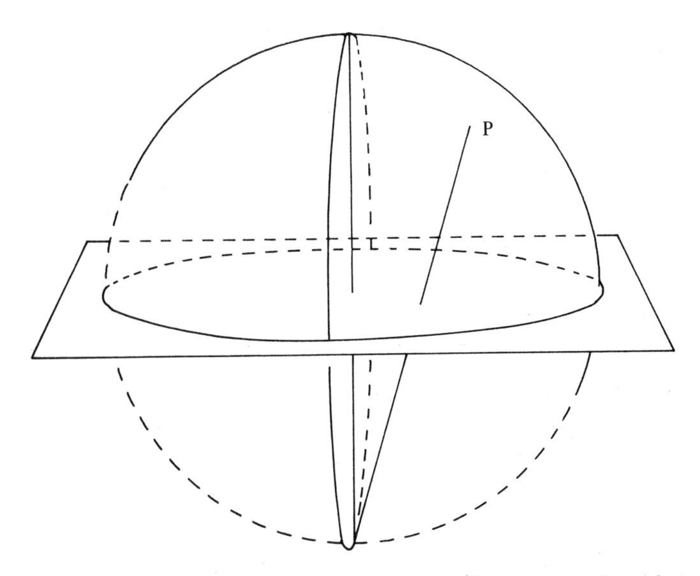

Fig. A.1. Stereographic projection of a point P, on the surface of a sphere onto a plane.

stereographic diagram of the faces of a cube viewed along its body diagonal.

The stereographic projection has two important and useful properties. First, all points that lie on circles, great or small, on the surface of the sphere also lie on circles in projection. Second, the angle between the

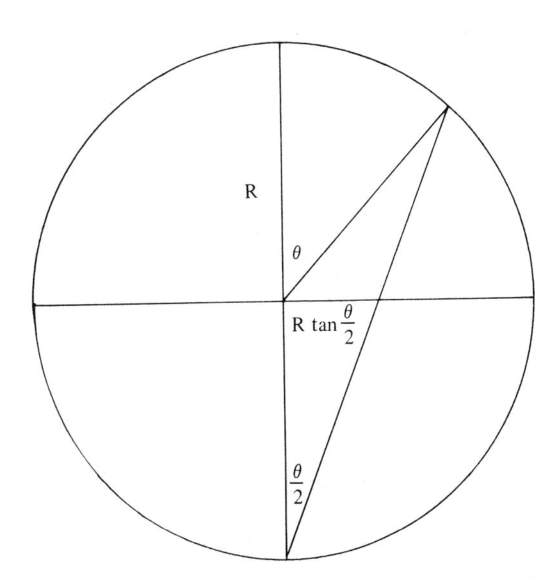

Fig. A.2. The construction used to determine the radial position of a point projected stereographically from a sphere.

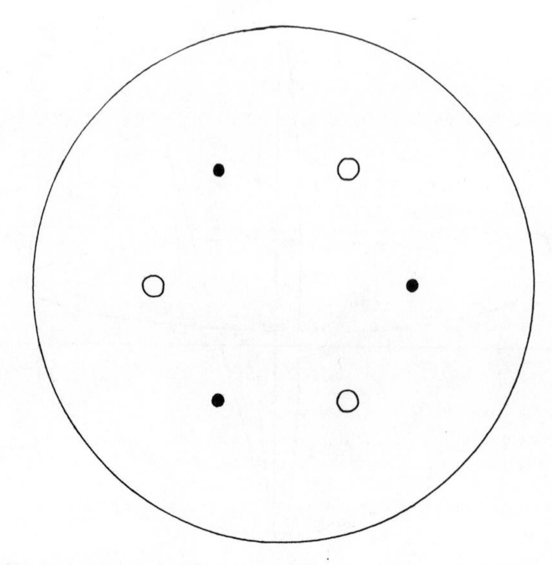

Fig. A.3. A stereographic projection of the faces of a cube, as seen looking down the body diagonal.

tangents to the projections of two great circles at their intersection is equal to the dihedral angle between the planes of the great circles, irrespective of where in the projection the intersection occurs. Thus, the pattern of the projections of circles that intersect in a symmetry axis reflects the symmetry of the axis.

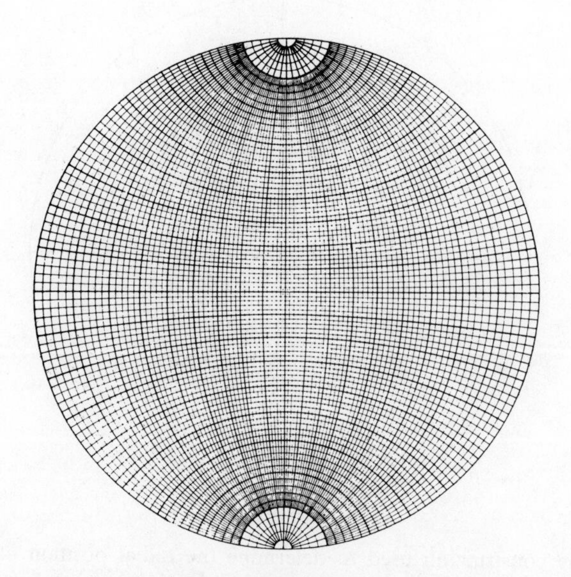

Fig. A.4. A stereographic, or Wulf, net showing the traces of great and small circles.

In practical applications of stereographic projection it is useful to have a chart of the projected traces of parallels of latitude and meridians of longitude, with the poles at the top and bottom, a so-called *Wulf net*. Wulf nets may be obtained from various commercial suppliers or, in this era of computer-driven plotters, one may be produced with a relatively trivial program. Figure A.4 is an example of such a chart. I have one that is 10 inches in diameter, mounted on a drawing board with a thumb tack sticking up through the middle. (It is important not to put your hand down on the point of the tack!) A sheet of translucent paper is laid on top of it, with the tack sticking through so that the paper may be rotated about the center. The radial coordinate represents the χ angle of a full-circle goniometer, and the angular coordinate represents the φ angle. Reflections found in a systematic search on a diffractometer are plotted on the paper. The angle between the normals of two sets of planes is found by rotating the paper so that the projections of both normals fall on the same meridian, and then counting the parallels between them. Prominent zones lie on great circles, and the angles between zones identify the symmetry of the axis that is perpendicular to both zone axes. Such a plot can be of immense help in determining the orientation of an unknown crystal.

Appendix C: Sublattices and Superlattices

We list here the matrices, **S**, that generate unique superlattices with unit cell volumes two, three, and four times the cell volume of the original lattice. The corresponding unique sublattices, with cell volumes one-half, one-third, and one-fourth of the cell volume of the original lattice, are generated by applying the same matrices to the reciprocal lattice of each lattice, and then forming the lattices that are reciprocal to the resulting superlattices.

For $|\mathbf{S}| = 2$, the unique matrices are

$$\begin{pmatrix} 2 & 0 & 0 \\ 0 & 1 & 0 \\ 0 & 0 & 1 \end{pmatrix}, \begin{pmatrix} 1 & 0 & 0 \\ 0 & 2 & 0 \\ 0 & 0 & 1 \end{pmatrix}, \begin{pmatrix} 1 & 0 & 0 \\ 0 & 1 & 0 \\ 0 & 0 & 2 \end{pmatrix}, \begin{pmatrix} 2 & 0 & 0 \\ 1 & 1 & 0 \\ 0 & 0 & 1 \end{pmatrix}, \begin{pmatrix} 2 & 0 & 0 \\ 0 & 1 & 0 \\ 1 & 0 & 1 \end{pmatrix},$$

$$\begin{pmatrix} 1 & 0 & 0 \\ 0 & 1 & 1 \\ 0 & 0 & 2 \end{pmatrix}, \begin{pmatrix} 1 & 1 & 0 \\ 0 & 1 & 1 \\ 1 & 0 & 1 \end{pmatrix}.$$

For $|\mathbf{S}| = 3$, the unique matrices are

$$\begin{pmatrix} 3 & 0 & 0 \\ 0 & 1 & 0 \\ 0 & 0 & 1 \end{pmatrix}, \begin{pmatrix} 1 & 0 & 0 \\ 0 & 3 & 0 \\ 0 & 0 & 1 \end{pmatrix}, \begin{pmatrix} 1 & 0 & 0 \\ 0 & 1 & 0 \\ 0 & 0 & 3 \end{pmatrix}, \begin{pmatrix} 1 & -1 & 0 \\ 2 & 1 & 0 \\ 0 & 0 & 1 \end{pmatrix}, \begin{pmatrix} 1 & 1 & 0 \\ -2 & 1 & 0 \\ 0 & 0 & 1 \end{pmatrix},$$

$$\begin{pmatrix} -1 & 0 & 1 \\ 2 & 0 & 1 \\ 0 & 1 & 0 \end{pmatrix}, \begin{pmatrix} 1 & 0 & 1 \\ 2 & 0 & -1 \\ 0 & 1 & 0 \end{pmatrix}, \begin{pmatrix} 0 & 1 & -1 \\ 0 & 2 & 1 \\ 1 & 0 & 0 \end{pmatrix}, \begin{pmatrix} 0 & 1 & 1 \\ 0 & -2 & 1 \\ 1 & 0 & 0 \end{pmatrix}, \begin{pmatrix} 2 & 1 & 1 \\ 1 & 1 & 0 \\ 0 & 2 & 1 \end{pmatrix},$$

$$\begin{pmatrix} 1 & 2 & 1 \\ -1 & -1 & 0 \\ 2 & 0 & 1 \end{pmatrix}, \begin{pmatrix} 1 & 1 & 2 \\ 1 & 0 & 1 \\ 2 & 1 & 0 \end{pmatrix}, \begin{pmatrix} 1 & 1 & 1 \\ 1 & 2 & 0 \\ 0 & 2 & 1 \end{pmatrix}.$$

For $|\mathbf{S}| = 4$, the unique matrices are

$$\begin{bmatrix}4&0&0\\0&1&0\\0&0&1\end{bmatrix}, \begin{bmatrix}1&0&0\\0&4&0\\0&0&1\end{bmatrix}, \begin{bmatrix}1&0&0\\0&1&0\\0&0&4\end{bmatrix}, \begin{bmatrix}4&0&0\\3&1&0\\0&0&1\end{bmatrix}, \begin{bmatrix}4&0&0\\1&1&0\\0&0&1\end{bmatrix},$$

$$\begin{bmatrix}4&0&0\\0&1&0\\3&0&1\end{bmatrix}, \begin{bmatrix}4&0&0\\0&1&0\\1&0&1\end{bmatrix}, \begin{bmatrix}1&0&0\\0&1&3\\0&0&4\end{bmatrix}, \begin{bmatrix}1&0&0\\0&1&1\\0&0&4\end{bmatrix}, \begin{bmatrix}4&0&0\\2&1&0\\0&0&1\end{bmatrix},$$

$$\begin{bmatrix}4&0&0\\0&1&0\\2&0&1\end{bmatrix}, \begin{bmatrix}1&0&0\\0&2&1\\0&0&2\end{bmatrix}, \begin{bmatrix}2&0&0\\1&2&0\\0&0&1\end{bmatrix}, \begin{bmatrix}2&0&0\\0&1&0\\1&0&2\end{bmatrix}, \begin{bmatrix}1&0&0\\0&1&2\\0&0&4\end{bmatrix},$$

$$\begin{bmatrix}2&2&0\\0&1&1\\1&0&1\end{bmatrix}, \begin{bmatrix}1&0&1\\0&1&1\\2&0&2\end{bmatrix}, \begin{bmatrix}1&1&0\\0&2&2\\1&0&1\end{bmatrix}, \begin{bmatrix}1&2&1\\1&1&2\\2&1&1\end{bmatrix}, \begin{bmatrix}3&1&0\\1&1&1\\2&0&1\end{bmatrix},$$

$$\begin{bmatrix}4&0&0\\1&1&0\\2&0&1\end{bmatrix}, \begin{bmatrix}2&1&0\\1&1&1\\3&0&1\end{bmatrix}, \begin{bmatrix}4&0&0\\2&1&0\\1&0&1\end{bmatrix}, \begin{bmatrix}1&1&1\\0&1&3\\1&0&2\end{bmatrix}, \begin{bmatrix}2&0&0\\0&1&1\\1&0&2\end{bmatrix},$$

$$\begin{bmatrix}2&1&0\\0&1&1\\2&0&1\end{bmatrix}, \begin{bmatrix}1&2&0\\0&2&1\\1&0&1\end{bmatrix}, \begin{bmatrix}1&1&0\\0&1&2\\1&0&1\end{bmatrix}, \begin{bmatrix}2&0&0\\0&2&0\\0&0&1\end{bmatrix}, \begin{bmatrix}2&0&0\\0&1&0\\0&0&2\end{bmatrix},$$

$$\begin{bmatrix}1&0&0\\0&2&0\\0&0&2\end{bmatrix}, \begin{bmatrix}2&0&0\\0&1&1\\0&0&2\end{bmatrix}, \begin{bmatrix}2&0&0\\0&2&0\\1&0&1\end{bmatrix}, \begin{bmatrix}2&0&0\\1&1&0\\0&0&2\end{bmatrix}, \begin{bmatrix}2&0&0\\1&1&1\\0&0&2\end{bmatrix}.$$

Appendix D: The Probability Integral, the Gamma Function, and Related Topics

We have made extensive use of probability density functions that included *normalization factors*, N, such that

$$(1/N)\int_{-\infty}^{+\infty} f(x)\,dx = 1.$$

In most cases we have stated without proof what these factors are, but in many cases it is useful to have some idea of their approximate magnitudes. The most frequently encountered of these is the normalization factor for the normal, or Gaussian, distribution. To determine this we need to evaluate the so-called probability integral,

$$P = \int_{-\infty}^{+\infty} \exp\left[-(1/2)x^2\right] dx.$$

Consider first the surface integral

$$\int_{-\infty}^{+\infty} dy \int_{-\infty}^{+\infty} \exp\left[-(1/2)(x^2 + y^2\right] dx,$$

and also the corresponding function in polar coordinates

$$\int_{-\pi}^{+\pi} d\theta \int_{0}^{+\infty} r \exp\left[-(1/2)r^2\right] dr.$$

Because the integrand is everywhere positive, we can write inequalities (see

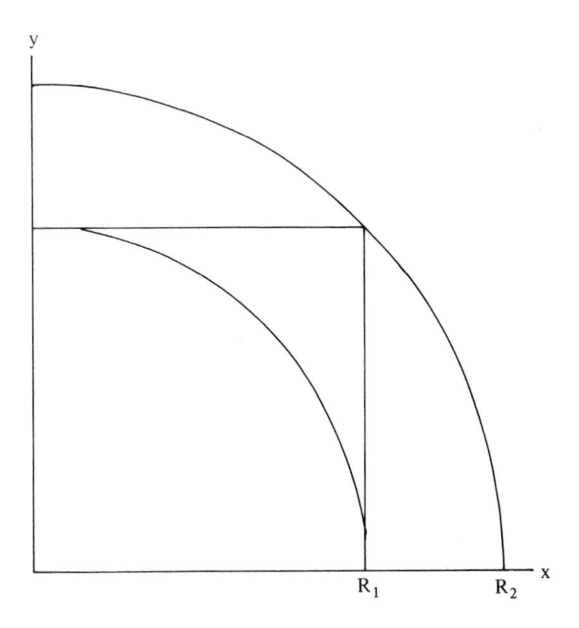

Fig. D.1. The regions over which the integrations are performed in evaluating the probability integral and the beta function.

Figure D.1) of the form

$$\int_{-\pi}^{+\pi} d\theta \int_{0}^{R_1} r \exp\left[-(1/2)r^2\right] dr \leqslant \int_{-R_1}^{+R_1} \exp\left[-(1/2)y^2\right] dy$$

$$\times \int_{-R_1}^{+R_1} \exp\left[-(1/2)x^2\right] dx$$

$$\leqslant \int_{-\pi}^{+\pi} d\theta \int_{0}^{R_2} r \exp\left[-(1/2)r^2\right] dr.$$

The first and last of these integrals can be evaluated in closed form, and the middle one is the product of two identical, definite integrals, so we have

$$2\pi\left\{1-\exp\left[-(1/2)R_1^2\right]\right\} \leqslant \left\{\int_{-R_1}^{+R_1} \exp\left[-(1/2)x^2\right] dx\right\}^2$$

$$\leqslant 2\pi\left\{1-\exp\left[-(1/2)R_2^2\right]\right\}.$$

Now, as R_1 and R_2 $[=\sqrt{2}\,R_1]$ become infinite, the first and last of these expressions approach 2π. The middle one, therefore, must also approach

2π, and

$$\int_{-\infty}^{+\infty} \exp\left[-(1/2)x^2\right] dx = (2\pi)^{1/2}.$$

We are also interested in the moments of the normal distribution, quantities of the form

$$\int_{-\infty}^{+\infty} x^n \exp\left[-(1/2)x^2\right] dx.$$

If n is odd, then the integral vanishes, because

$$\int_{-\infty}^{0} x^n \exp\left[-(1/2)x^2\right] dx = -\int_{0}^{+\infty} x^n \exp\left[-(1/2)x^2\right] dx.$$

If n is even, we can set $u(x) = x^{n-1}$ and $dv(x) = x\exp[-(1/2)x^2]dx$. Then, integrating by parts, we obtain

$$\int_{-\infty}^{+\infty} x^n \exp\left[-(1/2)x^2\right] dx = \left\{-x^{n-1} \exp\left[-(1/2)x^2\right]\right\}_{-\infty}^{+\infty}$$
$$+ (n-1)\int_{-\infty}^{+\infty} x^{n-2} \exp\left[-(1/2)x^2\right] dx.$$

The function inside the curly braces vanishes at both limits, so the quantity is equal to zero. We can repeat the process $(n/2 - 1)$ more times, getting to

$$\int_{-\infty}^{+\infty} x^n \exp\left[-(1/2)x^2\right] dx = (n-1)(n-3)\cdots 3\cdot 1\int_{-\infty}^{+\infty}\left[-(1/2)x^2\right] dx$$
$$= (n-1)(n-3)\cdots 3\cdot 1(2\pi)^{1/2}.$$

In probability density functions we frequently encounter the *gamma function*, designated $\Gamma(x)$. This function is defined by the relationship

$$\Gamma(x) = \int_{0}^{+\infty} \exp(-t)t^{x-1}\, dt.$$

Integrating by parts, this becomes

$$\Gamma(x) = \left[(1/x)t^x \exp(-t)\right]_0^{+\infty} + (1/x)\int_{0}^{+\infty} \exp(-t)t^x\, dt,$$

or $x\Gamma(x) = \Gamma(x+1)$.

$$\Gamma(1) = \int_{0}^{+\infty} \exp(-t)\, dt = 1.$$

It follows from these two relationships that $\Gamma(n) = (n-1)!$ when n is a positive integer. If $x = (1/2)$, we make the change of variable $t = y^2/2$,

$dt = y\,dy$, giving

$$\Gamma(1/2) = \sqrt{2}\int_0^{+\infty} \exp(-y^2/2)\,dy = \sqrt{\pi}\,.$$

From this we get $\Gamma(3/2) = \sqrt{\pi}/2$, $\Gamma(5/2) = 3\sqrt{\pi}/4$, etc. If, in the definition of the gamma function, we make the substitutions $x = \nu/2$ and $t = \chi^2/2$, we obtain

$$\Gamma(\nu/2) = 2^{-\nu/2}\int_0^{+\infty} \exp(-\chi^2/2)(\chi^2)^{\nu/2-1}\,d(\chi^2).$$

The integrand of the expression on the right is the density function for the χ^2 distribution, which confirms that $[2^{\nu/2}\Gamma(\nu/2)]^{-1}$ is the correct normalizing factor.

A function closely related to the gamma function is called the *beta function* and is designated $B(x, y)$. Its definition is

$$B(x, y) = \int_0^1 t^{x-1}(1-t)^{y-1}\,dt.$$

If we make the change of variable $t = \sin^2\theta$, we get one alternative definition

$$B(x, y) = 2\int_0^{\pi/2} (\sin\theta)^{2x-1}(\cos\theta)^{2y-1}\,d\theta.$$

The substitution $u = t/(1+t)$ gives another alternative definition

$$B(x, y) = \int_0^{+\infty} \left[u^{x-1}/(1+u)^{x+y}\right] du.$$

To see how it is related to the gamma function we consider the function $t^{2x-1}u^{2y-1}\exp(-t^2-u^2)$, and its equivalent in polar coordinates, $(\cos\theta)^{2x-1}(\sin\theta)^{2y-1}r^{2x+2y-2}\exp(-r^2)$. This function is nonnegative for t and $u \geqslant 0$ and for $0 \leqslant \theta \leqslant \pi/2$. We integrate the function over the regions shown in Figure D.1, giving the inequalities

$$\int_0^{\pi/2} (\cos\theta)^{2x-1}(\sin\theta)^{2y-1}\,d\theta \int_0^{R_1} \exp(-r^2)r^{2x+2y-1}\,dr$$

$$\leqslant \int_0^{R_1} t^{2x-1}\exp(-t^2)\,dt \int_0^{R_1} u^{2y-1}\exp(-u^2)\,du$$

$$\leqslant \int_0^{\pi/2} (\cos\theta)^{2x-1}(\sin\theta)^{2y-1}\,d\theta \int_0^{R_2} \exp(-r^2)r^{2x+2y-1}\,dr.$$

If we set $t^2 = z$ and $2t\,dt = dz$, then

$$\int_0^{+\infty} t^{2x-1}\exp(-t^2)\,dt = 2\int_0^{+\infty} z^{x-1}\exp(-z)\,dz = 2\Gamma(x).$$

By making similar substitutions, and allowing R_1 and R_2 to become infinite, we get $(1/2)B(x, y)(1/2)\Gamma(x + y) = [\Gamma(x)/2][\Gamma(y)/2]$, or $B(x, y) = [\Gamma(x) \cdot \Gamma(y)]/\Gamma(x + y)$. Setting $x = \nu_1/2$, $y = \nu_2/2$, and $u = (\nu_1/\nu_2)F$, and substituting in the third definition of the beta function, we obtain

$$B(\nu_1/2, \nu_2/2) = (\nu_1/\nu_2)^{\nu_1/2} \int_0^{\infty} F^{(\nu_1-2)/2}\left[1 + (\nu_1/\nu_2)F\right]^{-(\nu_1+\nu_2)/2} dF,$$

which confirms the normalizing factor for the F distribution.

We have repeatedly encountered the Fourier transform of the Gaussian density function, $G(x, \mu, \sigma) = (2\pi\sigma^2)^{-1/2}\exp\{-(1/2)[(x - \mu)/\sigma]^2\}$. The Fourier transform, or characteristic function, is

$$\Phi(t) = (2\pi\sigma^2)^{-1/2} \int_{-\infty}^{+\infty} \exp(ixt)\exp\left\{-(1/2)\left[(x - \mu)/\sigma\right]^2\right\} dx.$$

First we set $y = x - \mu$, $dy = dx$.

$$\Phi(t) = (2\pi\sigma^2)^{-1/2} \exp(i\mu t) \int_{-\infty}^{+\infty} \exp(iyt)\exp\left\{-(1/2)\left[y/\sigma\right]^2\right\} dy.$$

Now $\exp(iyt) = \cos(yt) + i\sin(yt)$. Because the other exponential in the integrand is an even function of y, the sine integral vanishes, leaving

$$f(t) = \int_{-\infty}^{+\infty} \cos(yt)\exp\left\{-(1/2)\left[y/\sigma\right]^2\right\} dy.$$

Differentiation of $f(t)$ with respect to t gives

$$\frac{d}{dt} f(t) = -\int_{-\infty}^{+\infty} y\sin(yt)\exp\left\{-(1/2)\left[y/\sigma\right]^2\right\} dy,$$

or, integrating by parts

$$\frac{d}{dt} f(t) = \left[\sigma^2 \sin(yt)\exp\left\{-(1/2)\left[y/\sigma\right]^2\right\}\right]_{-\infty}^{+\infty}$$

$$-\sigma^2 t \int_{-\infty}^{+\infty} \cos(yt)\exp\left\{-(1/2)\left[y/\sigma\right]^2\right\} dy.$$

The quantity in brackets vanishes at both limits, and the integral is identical to $f(t)$. Therefore $(d/dt)f(t) = -\sigma^2 t f(t)$. This is a differential equation whose solution is $\ln[f(t)] = -(1/2)\sigma t^2 + c$, or $f(t) = C\exp \cdot [-(1/2)\sigma^2 t^2]$. To evaluate the constant of integration, set $t = 0$, and $z = y/\sigma$, giving

$$f(0) = \sigma \int_{-\infty}^{+\infty} \exp\left[-(1/2)z^2\right] dz = (2\pi\sigma^2)^{1/2}.$$

Finally, $\Phi(t) = (2\pi\sigma^2)^{-1/2}\exp(i\mu t)f(t) = \exp[i\mu t - (1/2)\sigma^2 t^2]$.

Appendix E: The Harmonic Oscillator in Quantum Mechanics: Bloch's Theorem

Bloch's theorem, which states that the probability distribution for a particle in a harmonic potential well—that is, whose potential energy is a quadratic function of its displacement from some equilibrium position—is Gaussian, is so fundamental to crystallography that it seems worthwhile to take a look at the underlying mathematics. We shall first show that the Schrödinger equation can yield acceptable solutions only if the energy of the particle is equal to $[n + (1/2)]\hbar\omega_c$, where ω_c is the frequency of a classic harmonic oscillator with the same mass and potential function. Then we shall derive the form of the wave functions, ψ_n, and finally we shall show that the probability density function at any temperature is Gaussian.

Let us first examine the harmonic oscillator in one dimension. The basic postulate of quantum mechanics states that the probability density function for a particle with energy E in a region where the potential energy is described by a function, $V(x)$, is $\psi_E(x)\psi_E^*(x)$, where $\psi(x)$ is a solution to Schrödinger's equation, $(\hbar^2/2m)(d^2\psi/dx^2) + [E - V(x)]\psi(x) = 0$. Here $\hbar$ represents Planck's constant, h, divided by 2π, and m is the mass of the particle. $\psi(x)\psi^*(x)$ must be an acceptable probability density function, meaning that it must be everywhere greater than or equal to zero, and

$$\int_{-\infty}^{+\infty} \psi(x)\psi^*(x)\,dx = 1.$$

For the harmonic oscillator V is a quadratic function of x, and it is convenient to represent it by $V = a^2x^2$, where $a = (m\omega_c/\hbar)$. Making another convenient substitution, $\tau = 2mE/\hbar^2$, Schrödinger's equation can be written $(d^2\psi/dx^2) + (\tau - a^2x^2)\psi(x) = 0$. If $|x|$ is large enough, τ can be neglected relative to a^2x^2, and the equation reduces to $(d^2\psi/dx^2) = a^2x^2\psi(x)$. Because $(d/dx^2)[\exp(-ax^2/2] = (a^2x^2 - a)\exp(-ax^2/2)$, and, again, a may be neglected relative to a^2x^2 if $|x|$ is large enough, we expect $\psi(x)$ to behave like $\exp(-ax^2/2)$ in the limit of large $|x|$. We shall look, therefore, for solutions of the type $\psi(x) = Nf(x)\exp(-ax^2/2)$, where $f(x)$ satisfies the condition that

$$\int_{-\infty}^{+\infty} \left[f(x)\exp(-ax^2/2)\right]^2 dx = 1/N^2,$$

and N is a finite number greater than zero.

Substituting this trial function in Schrödinger's equation, it becomes

$$\exp(-ax^2/2)\left\{(\tau - a)f(x) - 2ax\frac{d}{dx}\left[f(x)\right] + \frac{d^2}{dx^2}\left[f(x)\right]\right\} = 0,$$

and, because the exponential factor is always greater than zero,

$$\left\{(\tau - a)f(x) - 2ax\frac{d}{dx}\left[f(x)\right] + \frac{d^2}{dx^2}\left[f(x)\right]\right\} = 0.$$

We assume that $f(x)$ can be represented by a polynomial of the type

$$f(x) = \sum_{k=0}^{n} c_k x^k.$$

Then

$$\frac{d}{dx}\left[f(x)\right] = \sum_{k=0}^{n-1} (k+1)c_{k+1}x^k, \quad \text{and}$$

$$\frac{d^2}{dx^2}\left[f(x)\right] = \sum_{k=0}^{n-2} (k+1)(k+2)c_{k+2}x^k.$$

For a polynomial to be equal to zero for all values of x, the coefficients of all powers of x must vanish individually, so that

$$(\tau - a - 2ak)c_k + (k+1)(k+2)c_{k+2} = 0,$$

or

$$c_{k+2} = -\left[(\tau - a - 2ak)c_k\right]/\left[(k+1)(k+2)\right].$$

This recursion relationship gives a form that ensures that any function, $f(x)$, that fits it, with arbitrary values of c_0 and c_1, will satisfy the differential equation, so we must examine the finite integral condition. First, if $(\tau - a - 2ak) = 0$ for some k, then $c_{k+2n} = 0$ for all $n \geqslant 1$. If this value of k is even, c_1 may be set equal to zero and, correspondingly, if k is odd, c_0 may be set equal to zero. In such a case $f(x)$ is a polynomial of finite degree, and

$$\int_{-\infty}^{+\infty} \left[f(x)\right]^2 \exp(-ax^2)\, dx \quad \text{clearly converges.}$$

If $(\tau - a - 2ak) \neq 0$ for any k, $f(x)$ is an infinite series if either c_0 or $c_1 \neq 0$. (The trivial solution $\psi(x) \equiv 0$ does not interest us.) It can be shown that this infinite series converges to a function that behaves like $\exp(2ax^2)$ at large x, and the required integral does not converge.

There is, therefore, no solution to Schrödinger's equation that satisfies the finite integral condition unless $(\tau - a - 2ak) = 0$ for some k. If this condition is satisfied for some k, then substituting for the values of τ and a we obtain $E = [k + (1/2)]\hbar\omega$. The general solutions have the form

$$\psi_n(x) = NH_n(x)\exp(-ax^2/2),$$

where

$$H_n(x) = \sum_{k=0}^{n} c_k x^k,$$

and all coefficients, c_k, where k has different parity from n are equal to zero. Thus $H_n(x)$ is a polynomial containing either odd or even powers of x, but not both. The functions $\psi_n(x)$ are the *wave functions* corresponding to allowed energy levels $E_n = [n + (1/2)]\hbar\omega_c$.

We shall now proceed to show that, except for an arbitrary scale factor, the polynomials $H_n(t)$, where $t = \sqrt{a}\, x$, can be generated by the procedure

$$H_n(t) = (-1)^n \exp(t^2) \frac{d^n}{dt^n}\left[\exp(-t^2)\right].$$

If this is true for some n greater than zero, then

$$H_{n+1}(t) = -\exp(t^2)\frac{d^{n+1}}{dt^{n+1}}\exp(-t^2), \quad \text{and}$$

$$\frac{d^{n+1}}{dt^{n+1}}\exp(-t^2) = \frac{d}{dt}\left[\frac{d^n}{dt^n}\exp(-t^2)\right] = \frac{d}{dt}\left[(-1)^n H_n(t)\exp(-t^2)\right].$$

Performing the indicated differentiation, we get

$$\frac{d^{n+1}}{dt^{n+1}}\exp(-t^2) = \left[\frac{d}{dt}H_n(t) - 2tH_n(t)\right]\exp(-t^2).$$

Therefore, if our hypothesis is correct,

$$H_{n+1}(t) = 2tH_n(t) - \frac{d}{dt}H_n(t). \tag{1}$$

Making use of the recursion formula, if n is even,

$$H_n(t) = c_0\left\{1 - \left[2nt^2/1\cdot 2\right] + \left[2^2 n(n-2)t^4/1\cdot 2\cdot 3\cdot 4\right] - \cdots\right\},$$

or, in general,

$$H_n(t) = c_0 \sum_{k=0}^{n/2} \{(-1)^k 2^{2k} t^{2k} (n/2)!\}/\{(2k)!(n/2-k)!\}, \tag{2}$$

Similarly, if n is odd,

$$H_n(t) = c_1 \sum_{k=0}^{(n-1)/2} \frac{(-1)^k 2^{2k} t^{2k+1} [(n-1)/2]!}{(2k+1)![(n-1)/2-k]!}. \tag{3}$$

If n is even, Equation (1) is applied to Equation (2), giving

$$H_{n+1}(t) = (n/2)!c_0 \left[(-1)^{n/2} 2^{n+1} t^{n+1}/n! + \sum_{k=1}^{n/2} \frac{[(-1)^{k-1} 2^{2k-1} t^{2k-1}]}{[(2k-2)!(n/2-k)!]} \right.$$
$$\left. \times \{1/(n/2-k+1) + 2/(2k+1)\} \right].$$

If we now set $m = n + 1$, set $c_1 = 2(n+1)c_0$, and simplify, we get

$$H_m(t) = [(m-1)/2]!c_1$$
$$\times \sum_{k=0}^{(m-1)/2} \{(-1)^k 2^{2k} t^{2k+1}\}/\{(2k+1)![(m-1)/2-k]!\},$$

which agrees with Equation (3). If n is odd, Equation (1) is applied to Equation (3), giving

$$H_{n+1}(t) = [(n-1)/2]!c_1 \left[1/\{(n-1)/2\}! + (-1)^{(n-1)/2} 2^n t^{n+1}/n! \right.$$
$$+ \sum_{k=1}^{(n-1)/2} \frac{[(-1)^{k-1} 2^{2k-1} t^{2k}]}{[(2k-1)!\{(n-1)/2-k\}!]}$$
$$\left. \times \{1/[(n-1)/2-k+1] + 2/2k\} \right].$$

Again set $m = n + 1$, set $c_0 = -c_1$, and simplify, obtaining

$$H_n(t) = (m/2)!c_0 \sum_{k=0}^{m/2} [(-1)^k 2^{2k} t^{2k}]/[(2k)!(m/2-k)!],$$

which agrees with Equation (2). Applying Equation (2) for $n = 0$ gives $H_0(t) = c_0$. Applying Equation (1) to this gives $H_1(t) = 2tc_0$, which agrees with Equation (3) for $n = 1$. Equation (1) is therefore established by induction.

The polynomials $H_n(t)$ are the *other* set of Hermite polynomials. The first few polynomials in *this* set are

$$H_0(t) = 1,$$

$$H_1(t) = 2t,$$

$$H_2(t) = 4t^2 - 2,$$

$$H_3(t) = 8t^3 - 12t,$$

$$H_4(t) = 16t^4 - 48t^2 + 12.$$

These polynomials are quite distinct from the ones that are used in the Edgeworth series for approximating statistical distributions (see Chapter 9), and it is important to be sure which set is in use at a given time!

Following a procedure similar to the one we used in Chapter 9, we shall derive several useful relationships among these Hermite polynomials. First, consider the function $\gamma(x) = \exp(-x^2)$, $\gamma(x - t) = \exp(-x^2 + 2xt - t^2)$, or $\gamma(x - t) = \gamma(x)\exp(2xt - t^2)$. If we expand $\gamma(x - t)$ in a Taylor's series in powers of t we obtain

$$\gamma(x - t) = \sum_{j=0}^{\infty} \left[-t^j/j!\right] \frac{d^j}{dx^j} \gamma(x),$$

or, from the definition of $H_j(x)$,

$$\gamma(x - t) = \gamma(x) \sum_{j=0}^{\infty} \left[t^j/j!\right] H_j(x).$$

We therefore have the identity

$$\exp(2tx - t^2) \equiv \sum_{j=0}^{\infty} \left[t^j/j!\right] H_j(x).$$

Differentiating both sides of the identity with respect to t, we get

$$2(x - t)\exp(2xt - t^2) \equiv \sum_{j=0}^{\infty} t^{j-1} H_j(x)/(j-1)!,$$

or

$$2x \sum_{j=0}^{\infty} t^j H_j(x)/j! - 2 \sum_{j=0}^{\infty} t^{j+1} H_j(x)/j! \equiv \sum_{j=0}^{\infty} t^{j-1} H_j(x)/(j-1)!.$$

For the expressions on both sides to be identical, the coefficients of each power of t must be individually identical, from which it follows that

$$2xH_r(x)/r! - 2H_{r-1}(x)/(r-1)! = H_{r+1}(x)/r!,$$

or

$$2xH_r(x) - 2rH_{r-1}(x) = H_{r+1}(x).$$

Next we differentiate both sides of the identity with respect to x, giving

$$2t\exp(2xt - t^2) = \sum_{j=0}^{\infty}(t^j/j!)\frac{d}{dx}H_j(x),$$

or

$$2\sum_{j=0}^{\infty}(t^{j+1}/j!)H_j(x) = \sum_{j=0}^{\infty}(t^j/j!)\frac{d}{dx}H_j(x).$$

Again we equate coefficients of like powers of t, giving

$$\frac{d}{dx}H_r(x) = 2rH_{r-1}(x) = 2xH_r(x) - H_{r+1}(x).$$

We now consider the integral

$$\int_{-\infty}^{+\infty}H_m(x)H_n(x)\exp(-x^2)\,dx = 2^m!\int_{-\infty}^{+\infty}H_0(x)H_{n-m}(x)\exp(-x^2)\,dx.$$

Assuming $m \leqslant n$, we set $u = H_m(x)$, $dv = H_n(x)\exp(-x^2)\,dx$, and evaluate the integral on the right by parts, giving

$$\int_{-\infty}^{+\infty}H_m(x)H_n(x)\exp(-x^2)\,dx$$
$$= \left[-H_m(x)H_{n-1}(x)\exp(-x^2)\right]_{-\infty}^{+\infty}$$
$$+ 2m\int_{-\infty}^{+\infty}H_{m-1}(x)H_{n-1}(x)\exp(-x^2)\,dx.$$

The quantity in square brackets vanishes at both limits. If we repeat this process $(m-1)$ more times, we will get to

$$\int_{-\infty}^{+\infty}H_m(x)H_n(x)\exp(-x^2)\,dx = 2^m!\int_{-\infty}^{+\infty}H_0(x)H_{n-m}(x)\exp(-x^2)\,dx.$$

If $m \neq n$, the integral on the right vanishes, because the next step will involve $(d/dx)H_0(x) = 0$. If $m = n$, the integral becomes

$$\int_{-\infty}^{+\infty}\left[H_0(x)\right]^2\exp(-x^2)\,dx = \int_{-\infty}^{+\infty}\exp(-x^2)\,dx = \sqrt{\pi}\,.$$

Therefore

$$\int_{-\infty}^{+\infty}[H_n(x)]^2\exp(-x^2)\,dx = 2^n\sqrt{\pi}\,n! = 1/N_n^2.$$

In consequence we can now express the allowed wave functions for the harmonic oscillator in the form

$$\psi_n(x) = \{(\pi/a)^{1/2}2^n n!\}^{-1/2}H_n(\sqrt{a}\,x)\exp(-ax^2/2).$$

Using this result, we shall now determine the probability distribution for a particle in a harmonic potential well at a finite temperature. If the system is in thermal equilibrium, the probability of finding the particle in the state whose wave function is ψ_n is proportional to a Boltzmann weighting factor, $\exp(-E_n/kT)$. The total probability density at the point t is therefore

$$\Phi(t) = (1/Z)\sum_{n=0}^{+\infty}[\psi_n(t)]^2\exp\{-[n+(1/2)]\hbar\omega_c/kT\},$$

where

$$Z = \sum_{n=0}^{\infty}\exp\{-[n+(1/2)]\hbar\omega_c/kT\}$$

$$= \exp(-\hbar\omega_c/2kT)/[1+\exp(\hbar\omega_c/kT)].$$

Noting that $H_n(t) = (\sqrt{\pi}\,2^n n!)^{1/2}\exp(t^2/2)\psi_n(t)$, and that, from Equation (1), $(d/dt)H_n(t) = 2tH_n(t) - H_{n+1}(t)$, we can derive the two relationships

$$t\psi(t) = 2^{-1/2}[n^{1/2}\psi_{n-1}(t) + (n+1)^{1/2}\psi_{n+1}(t)] \quad \text{and}$$

$$\frac{d\psi}{dt} = 2^{-1/2}[n^{1/2}\psi_{n-1}(t) - (n+1)^{1/2}\psi_{n+1}(t)].$$

We can then rewrite the expression for $\Phi(t)$ in the form

$$\Phi(t) = (\sqrt{2}\,tZ)^{-1}\sum_{n=0}^{+\infty}\exp\{-[n+(1/2)]\hbar\omega_c/kT\}\psi_n(t)$$

$$\times[n^{1/2}\psi_{n-1}(t) + (n+1)^{1/2}\psi_{n+1}(t)].$$

In this expression each term in $\psi_n\psi_{n-1}$ appears twice, once multiplied by $-n^{1/2}\exp\{-[n+(1/2)]\hbar\omega_c/kT\}$, and once multiplied by $-n^{1/2}\exp\{-[n+(3/2)]\hbar\omega_c/kT\}$, which is equal to $-n^{1/2}\exp(-\hbar\omega_c/kT)$

$\exp\{-[n+(1/2)]\hbar\omega_c/kT\}$, so that the expression for $\Phi(t)$ can be rearranged to give

$$\Phi(t) = \left(\sqrt{2}\,tZ\right)^{-1}\left[1+\exp(-\hbar\omega_c/kT)\right]$$

$$\times \sum_{n=1}^{+\infty} n^{1/2}\exp\left\{-\left[n+(1/2)\right]\hbar\omega_c/kT\right\}\psi_n(t)\psi_{n-1}(t).$$

In a similar manner, we can obtain

$$\frac{d\Phi}{dt} = (2/Z)\sum_{n=0}^{+\infty}\exp\left\{-\left[n+(1/2)\right]\hbar\omega_c/kT\right\}\psi_n(t)\frac{d\psi_n(t)}{dt},$$

which can be rewritten, using the expression for $(d\psi_n(t)/dt)$,

$$\frac{d\Phi(t)}{dt} = -\left(\sqrt{2}/Z\right)\left[1-\exp(\hbar\omega_c/kT)\right]$$

$$\times \sum_{n=1}^{+\infty} n^{1/2}\exp\left\{-\left[n+(1/2)\right]\hbar\omega_c/kT\right\}\psi_n(t)\psi_{n-1}(t).$$

If we set $\sigma^2 = (1/2)\coth[(1/2)h\omega_t/kT]$, and divide the expression for $(d\Phi(t)/dt)$ by the expression for $\Phi(t)$, we obtain $(d\Phi(t)/dt)/\Phi(t) = -t/\sigma^2$, a differential equation for which the solution is $\ln[\Phi(t) = -t^2/2\sigma^2 + C$, or $\Phi(t) = D\exp(-t^2/2\sigma^2)$, where $D = \exp(C)$. Because the probability of finding the particle somewhere must be unity, $D = (2\pi\sigma^2)^{-1/2}$.

This result proves Bloch's theorem in one dimension. In three dimensions the axes can always be chosen so that $V(\mathbf{x}) = a_1^2x_1^2 + a_2^2x_2^2 + a_3^2x_3^2$, and the wave functions can be expressed as products of one-dimensional wave functions, so that the probability density for a particle in a three-dimensional, harmonic, potential well is also Gaussian.

Appendix F: Symmetry Restrictions on Second-, Third-, and Fourth-Rank Tensors

Tensors of various ranks have restrictions on their elements imposed by point-group symmetry. Because the patterns are quite different for different ranks, we shall list each rank separately.

For second-rank tensors there is a separate set of restrictions for the rigid-body **S** tensor, because of its property of having partially polar and partially axial character.

Point group	Polar of axial tensors			**S** tensor		
	T_{11}	T_{12}	T_{13}	S_{11}	S_{12}	S_{13}
1	T_{12}	T_{22}	T_{23}	S_{21}	S_{22}	S_{23}
	T_{13}	T_{23}	T_{33}	S_{31}	S_{32}	S_{33}
	T_{11}	T_{12}	T_{13}			
$\bar{1}$	T_{12}	T_{22}	T_{23}		0	
	T_{13}	T_{23}	T_{33}			
	T_{11}	T_{12}	0	S_{11}	S_{12}	0
2 (z axis unique)	T_{12}	T_{22}	0	S_{21}	S_{22}	0
	0	0	T_{33}	0	0	S_{33}
	T_{11}	T_{12}	0	0	0	S_{13}
m	T_{12}	T_{22}	0	0	0	S_{23}
	0	0	T_{33}	S_{31}	S_{32}	0
	T_{11}	T_{12}	0			
$2/m$	T_{12}	T_{22}	0		0	
	0	0	T_{33}			
	T_{11}	0	0	S_{11}	0	0
222	0	T_{22}	0	0	S_{22}	0
	0	0	T_{33}	0	0	S_{33}

Point group	Polar of axial tensors			S tensor		
	T_{11}	0	0	0	S_{12}	0
$mm2$	0	T_{22}	0	S_{21}	0	0
	0	0	T_{33}	0	0	0
	T_{11}	0	0			
mmm	0	T_{22}	0		0	
	0	0	T_{33}			
	T_{11}	0	0	S_{11}	S_{12}	0
$3, 4, 6$	0	T_{11}	0	$-S_{12}$	S_{11}	0
	0	0	T_{33}	0	0	S_{33}
	T_{11}	0	0	S_{11}	S_{12}	0
$\bar{4}$	0	T_{11}	0	S_{12}	$-S_{11}$	0
	0	0	T_{33}	0	0	S_{33}
$\bar{3}, 4/m, \bar{6}$	T_{11}	0	0			
$6/m, 4/mmm, \bar{3}m,$	0	T_{11}	0		0	
$\bar{6}m2, 6/mmm$	0	0	T_{33}			
	T_{11}	0	0	S_{11}	0	0
$32, 422, 622$	0	T_{11}	0	0	S_{11}	0
	0	0	T_{33}	0	0	S_{33}
	T_{11}	0	0	0	S_{12}	0
$3m, 4mm, 6mm$	0	T_{11}	0	$-S_{12}$	0	0
	0	0	T_{33}	0	0	0
	T_{11}	0	0	S_{11}	0	0
$\bar{4}2m$	0	T_{11}	0	0	$-S_{11}$	0
	0	0	T_{33}	0	0	0
	T_{11}	0	0	S_{11}	0	0
$23, 432$	0	T_{11}	0	0	S_{11}	0
	0	0	T_{11}	0	0	S_{11}
	T_{11}	0	0			
$m3, \bar{4}3m, m3m$	0	T_{11}	0		0	
	0	0	T_{11}			

All elements of third-rank tensors vanish in the 11 centrosymmetric point groups, and also in point group 432. In the remaining point groups the allowed tensor elements are given in the scheme that applies to piezoelectric moduli, which give the electric dipole moment induced as a result of the application of a stress. In this case the tensor is symmetric under interchange of the second and third indices, but not under interchange of the first index with the second or the third. In the case of piezoelectric constants, which give the components of strain resulting from the application of an electric field, the tensor would be symmetric under interchange of the first two indices. Third moments and third cumulants are symmetric under all permutations of indices, which leads to additional restrictions.

There are no restrictions in point group 1. In point group 2 the tensor elements are

$$\begin{matrix} 0 & 0 & 0 & Q_{123} & Q_{113} & 0 \\ 0 & 0 & 0 & Q_{223} & Q_{213} & 0 \\ Q_{311} & Q_{322} & Q_{333} & 0 & 0 & Q_{312}. \end{matrix}$$

In point group m the elements are

$$\begin{matrix} Q_{111} & Q_{122} & Q_{133} & 0 & 0 & Q_{112} \\ Q_{211} & Q_{222} & Q_{233} & 0 & 0 & Q_{212} \\ 0 & 0 & 0 & Q_{323} & Q_{313} & 0. \end{matrix}$$

In point group 222 the elements are

$$\begin{matrix} 0 & 0 & 0 & Q_{123} & 0 & 0 \\ 0 & 0 & 0 & 0 & Q_{213} & 0 \\ 0 & 0 & 0 & 0 & 0 & Q_{312}. \end{matrix}$$

In point group $mm2$ the elements are

$$\begin{matrix} 0 & 0 & 0 & 0 & Q_{113} & 0 \\ 0 & 0 & 0 & Q_{223} & 0 & 0 \\ Q_{311} & Q_{322} & Q_{333} & 0 & 0 & 0. \end{matrix}$$

The effects of rotation axes of order 4 and higher are identical, giving several pairs of groups with the same patterns of elements. In point groups 4 and 6 the elements are

$$\begin{matrix} 0 & 0 & 0 & Q_{123} & Q_{113} & 0 \\ 0 & 0 & 0 & Q_{113} & -Q_{123} & 0 \\ Q_{311} & Q_{311} & Q_{333} & 0 & 0 & 0. \end{matrix}$$

Because moments and cumulants are fully symmetric ${}^3\mu_{123} = {}^3\mu_{213} = 0$. In point groups 422 and 622 the elements are

$$\begin{matrix} 0 & 0 & 0 & Q_{123} & 0 & 0 \\ 0 & 0 & 0 & 0 & -Q_{123} & 0 \\ 0 & 0 & 0 & 0 & 0 & 0. \end{matrix}$$

All third moments and cumulants vanish. In point groups $4mm$ and $6mm$ the elements are

$$\begin{matrix} 0 & 0 & 0 & 0 & Q_{113} & 0 \\ 0 & 0 & 0 & Q_{113} & 0 & 0 \\ Q_{311} & Q_{311} & Q_{333} & 0 & 0 & 0. \end{matrix}$$

In point group $\bar{4}$ the elements are

$$\begin{matrix} 0 & 0 & 0 & Q_{123} & Q_{113} & 0 \\ 0 & 0 & 0 & -Q_{113} & Q_{123} & 0 \\ Q_{311} & -Q_{311} & 0 & 0 & 0 & Q_{312}. \end{matrix}$$

In point group $\bar{4}2m$ the elements are

$$\begin{matrix} 0 & 0 & 0 & Q_{123} & 0 & 0 \\ 0 & 0 & 0 & 0 & Q_{123} & 0 \\ 0 & 0 & 0 & 0 & 0 & Q_{312}. \end{matrix}$$

In point group 3 the elements are

$$\begin{matrix} Q_{111} & -Q_{111} & 0 & Q_{123} & Q_{113} & -Q_{222} \\ -Q_{222} & Q_{222} & 0 & Q_{113} & -Q_{123} & -Q_{111} \\ Q_{311} & Q_{311} & Q_{333} & 0 & 0 & 0. \end{matrix}$$

In point group 32 the elements are

$$\begin{matrix} Q_{111} & -Q_{111} & 0 & Q_{123} & 0 & 0 \\ 0 & 0 & 0 & 0 & -Q_{123} & -Q_{111} \\ 0 & 0 & 0 & 0 & 0 & 0. \end{matrix}$$

In point group $3m$ the elements are

$$\begin{matrix} Q_{111} & -Q_{111} & 0 & 0 & Q_{113} & 0 \\ 0 & 0 & 0 & Q_{113} & 0 & -Q_{111} \\ Q_{311} & Q_{311} & Q_{333} & 0 & 0 & 0. \end{matrix}$$

In point group $\bar{6}$ the elements are

$$\begin{matrix} Q_{111} & -Q_{111} & 0 & 0 & 0 & -Q_{222} \\ -Q_{222} & Q_{222} & 0 & 0 & 0 & -Q_{111} \\ 0 & 0 & 0 & 0 & 0 & 0. \end{matrix}$$

In point group $\bar{6}m2$ the elements are

$$\begin{array}{cccccc} Q_{111} & -Q_{111} & 0 & 0 & 0 & 0 \\ 0 & 0 & 0 & 0 & 0 & -Q_{111} \\ 0 & 0 & 0 & 0 & 0 & 0. \end{array}$$

In point groups 23 and $\bar{4}3m$ the elements are

$$\begin{array}{cccccc} 0 & 0 & 0 & Q_{123} & 0 & 0 \\ 0 & 0 & 0 & 0 & Q_{123} & 0 \\ 0 & 0 & 0 & 0 & 0 & Q_{123}. \end{array}$$

The symmetry restrictions on fourth-rank tensor elements are given as they apply to elastic constants, for which $C_{ijkl} \neq C_{ikjl}$. For fourth moments and fourth cumulants, which are fully symmetric, there are further symmetry restrictions. These restrictions reduce the number of independent elements from 21 to 15 in the triclinic system, and from 3 to 2 in the cubic system. There are no symmetry-imposed restrictions in the triclinic point groups. In the monoclinic system, with the unique axis assumed to be y, the tensor elements are

$$\begin{array}{cccccc} C_{1111} & C_{1122} & C_{1133} & 0 & C_{1113} & 0 \\ & C_{2222} & C_{2233} & 0 & C_{2213} & 0 \\ & & C_{3333} & 0 & C_{3313} & 0 \\ & & & C_{2323} & 0 & C_{2312} \\ & & & & C_{1313} & 0 \\ & & & & & C_{1212}. \end{array}$$

In the orthorhombic system the tensor elements are

$$\begin{array}{cccccc} C_{1111} & C_{1122} & C_{1133} & 0 & 0 & 0 \\ & C_{2222} & C_{2233} & 0 & 0 & 0 \\ & & C_{3333} & 0 & 0 & 0 \\ & & & C_{2323} & 0 & 0 \\ & & & & C_{1313} & 0 \\ & & & & & C_{1212}. \end{array}$$

In the trigonal point groups 3 and $\bar{3}$ the allowed elements are

$$\begin{array}{cccccc} C_{1111} & C_{1122} & C_{1133} & C_{1123} & -C_{2213} & 0 \\ & C_{1111} & C_{1133} & -C_{1123} & C_{2213} & 0 \\ & & C_{3333} & 0 & 0 & 0 \\ & & & C_{2323} & 0 & C_{2213} \\ & & & & C_{2323} & C_{1123} \\ & & & & & (1/2)(C_{1111} - C_{1122}). \end{array}$$

In the trigonal point groups 32, $3m$, and $\bar{3}m$ the elements are

$$\begin{array}{cccccc}
C_{1111} & C_{1122} & C_{1133} & C_{1123} & 0 & 0 \\
 & C_{1111} & C_{1133} & -C_{1123} & 0 & 0 \\
 & & C_{3333} & 0 & 0 & 0 \\
 & & & C_{2323} & 0 & 0 \\
 & & & & C_{2323} & C_{1123} \\
 & & & & & (1/2)(C_{1111} - C_{1122}).
\end{array}$$

In the tetragonal system the elements are

$$\begin{array}{cccccc}
C_{1111} & C_{1122} & C_{1133} & 0 & 0 & 0 \\
 & C_{1111} & C_{1133} & 0 & 0 & 0 \\
 & & C_{3333} & 0 & 0 & 0 \\
 & & & C_{2323} & 0 & 0 \\
 & & & & C_{2323} & 0 \\
 & & & & & C_{1212}.
\end{array}$$

In the hexagonal system the elements are

$$\begin{array}{cccccc}
C_{1111} & C_{1122} & C_{1133} & 0 & 0 & 0 \\
 & C_{1111} & C_{1133} & 0 & 0 & 0 \\
 & & C_{3333} & 0 & 0 & 0 \\
 & & & C_{2323} & 0 & 0 \\
 & & & & C_{2323} & 0 \\
 & & & & & (1/2)(C_{1111} - C_{1122}).
\end{array}$$

In the cubic system the elements are

$$\begin{array}{cccccc}
C_{1111} & C_{1122} & C_{1122} & 0 & 0 & 0 \\
 & C_{1111} & C_{1122} & 0 & 0 & 0 \\
 & & C_{1111} & 0 & 0 & 0 \\
 & & & C_{2323} & 0 & 0 \\
 & & & & C_{2323} & 0 \\
 & & & & & C_{2323}.
\end{array}$$

Appendix G: Some Useful Computer Programs

On the following pages we reproduce FORTRAN source code for a number of useful mathematical operations, including the important statistical distribution functions. Also included are code for the libration correction—the function EIJ is the matrix $(\mathbf{I} + \mathbf{M})$ discussed in Chapter 9—and for shape and rigid-body-motion constraints. I am indebted to James J. Filliben for permission to use the statistical functions.

```
      REAL FUNCTION CHSCDF(X,NU)
C
C     PURPOSE--THIS SUBROUTINE COMPUTES THE CUMULATIVE DISTRIBUTION
C              FUNCTION VALUE FOR THE CHI-SQUARED DISTRIBUTION
C              WITH INTEGER DEGREES OF FREEDOM PARAMETER = NU.
C              THIS DISTRIBUTION IS DEFINED FOR ALL NON-NEGATIVE X.
C              THE PROBABILITY DENSITY FUNCTION IS GIVEN
C              IN THE REFERENCES BELOW.
C     INPUT  ARGUMENTS--X      = THE SINGLE PRECISION VALUE AT
C                                WHICH THE CUMULATIVE DISTRIBUTION
C                                FUNCTION IS TO BE EVALUATED.
C                                X SHOULD BE NON-NEGATIVE.
C                     --NU     = THE INTEGER NUMBER OF DEGREES
C                                OF FREEDOM.
C                                NU SHOULD BE POSITIVE.
C     OUTPUT--THE SINGLE PRECISION CUMULATIVE DISTRIBUTION
C             FUNCTION VALUE CDF FOR THE CHI-SQUARED DISTRIBUTION
C             WITH DEGREES OF FREEDOM PARAMETER = NU.
C     PRINTING--NONE UNLESS AN INPUT ARGUMENT ERROR CONDITION EXISTS.
C     RESTRICTIONS--X SHOULD BE NON-NEGATIVE.
C                 --NU SHOULD BE A POSITIVE INTEGER VARIABLE.
C     OTHER DATAPAC   SUBROUTINES NEEDED--GAUCDF.
C     FORTRAN LIBRARY SUBROUTINES NEEDED--DSQRT, DEXP.
C     MODE OF INTERNAL OPERATIONS--DOUBLE PRECISION.
C     LANGUAGE--ANSI FORTRAN.
C     REFERENCES--NATIONAL BUREAU OF STANDARDS APPLIED MATHEMATICS
C                 SERIES 55, 1964, PAGE 941, FORMULAE 26.4.4 AND 26.4.5.
C               --JOHNSON AND KOTZ, CONTINUOUS UNIVARIATE
C                 DISTRIBUTIONS--1, 1970, PAGE 176,
C                 FORMULA 28, AND PAGE 180, FORMULA 33.1.
C               --OWEN, HANDBOOK OF STATISTICAL TABLES,
C                 1962, PAGES 50-55.
C               --PEARSON AND HARTLEY, BIOMETRIKA TABLES
C                 FOR STATISTICIANS, VOLUME 1, 1954,
C                 PAGES 122-131.
C     WRITTEN BY--JAMES J. FILLIBEN
C                 STATISTICAL ENGINEERING LABORATORY (205.03)
C                 NATIONAL BUREAU OF STANDARDS
C                 WASHINGTON, D. C. 20234
C                 PHONE:  301-921-2315
C     ORIGINAL VERSION--JUNE      1972.
C     UPDATED         --MAY       1974.
C     UPDATED         --SEPTEMBER 1975.
C     UPDATED         --NOVEMBER  1975.
C     UPDATED         --OCTOBER   1976.
C     CONVERTED TO REAL FUNCTION BY E. PRINCE, JULY 18, 1980.
C
C---------------------------------------------------------------------
C
      DOUBLE PRECISION DX,PI,CHI,SUM,TERM,AI,DCDFN
      DOUBLE PRECISION DNU
      DOUBLE PRECISION DSQRT,DEXP
      DOUBLE PRECISION DLOG
      DOUBLE PRECISION DFACT,DPOWER
      DOUBLE PRECISION DW
      DOUBLE PRECISION D1,D2,D3
      DOUBLE PRECISION TERM0,TERM1,TERM2,TERM3,TERM4
      DOUBLE PRECISION B11
```

```
      DOUBLE PRECISION B21
      DOUBLE PRECISION B31,B32
      DOUBLE PRECISION B41,B42,B43
      DATA NUCUT/1000/
      DATA PI/3.14159265358979D0/
      DATA DPOWER/0.33333333333333D0/
      DATA B11/0.33333333333333D0/
      DATA B21/-0.02777777777778D0/
      DATA B31/-0.00061728395061D0/
      DATA B32/-13.0D0/
      DATA B41/0.00018004115226D0/
      DATA B42/6.0D0/
      DATA B43/17.0D0/
      DATA IPR/6/
C
C     CHECK THE INPUT ARGUMENTS FOR ERRORS
C
      IF(NU.LE.0)GOTO50
      IF(X.LT.0.0)GOTO55
      GOTO90
   50 WRITE(IPR,15)
      WRITE(IPR,47)NU
      CHSCDF=0.0
      RETURN
   55 WRITE(IPR,4)
      WRITE(IPR,46)X
      CHSCDF=0.0
      RETURN
   90 CONTINUE
    4 FORMAT(1H , 96H***** NON-FATAL DIAGNOSTIC--THE FIRST  INPUT ARGUME
     1NT TO THE CHSCDF SUBROUTINE IS NEGATIVE *****)
   15 FORMAT(1H , 91H***** FATAL ERROR--THE SECOND INPUT ARGUMENT TO THE
     1 CHSCDF SUBROUTINE IS NON-POSITIVE *****)
   46 FORMAT(1H , 35H***** THE VALUE OF THE ARGUMENT IS ,E15.8,6H *****)
   47 FORMAT(1H , 35H***** THE VALUE OF THE ARGUMENT IS ,I8   ,6H *****)
C
C-----START POINT-----------------------------------------------------
C
      DX=X
      ANU=NU
      DNU=NU
C
C     IF X IS NON-POSITIVE, SET CDF = 0.0 AND RETURN.
C     IF NU IS SMALLER THAN 10 AND X IS MORE THAN 200
C     STANDARD DEVIATIONS BELOW THE MEAN,
C     SET CDF = 0.0 AND RETURN.
C     IF NU IS 10 OR LARGER AND X IS MORE THAN 100
C     STANDARD DEVIATIONS BELOW THE MEAN,
C     SET CDF = 0.0 AND RETURN.
C     IF NU IS SMALLER THAN 10 AND X IS MORE THAN 200
C     STANDARD DEVIATIONS ABOVE THE MEAN,
C     SET CDF = 1.0 AND RETURN.
C     IF NU IS 10 OR LARGER AND X IS MORE THAN 100
C     STANDARD DEVIATIONS ABOVE THE MEAN,
C     SET CDF = 1.0 AND RETURN.
C
      IF(X.LE.0.0)GOTO105
      AMEAN=ANU
      SD=SQRT(2.0*ANU)
```

```
      Z=(X-AMEAN)/SD
      IF(NU.LT.10.AND.Z.LT.-200.0)GOTO105
      IF(NU.GE.10.AND.Z.LT.-100.0)GOTO105
      IF(NU.LT.10.AND.Z.GT.200.0)GOTO107
      IF(NU.GE.10.AND.Z.GT.100.0)GOTO107
      GOTO109
  105 CHSCDF=0.0
      RETURN
  107 CHSCDF=1.0
      RETURN
  109 CONTINUE
C
C     DISTINGUISH BETWEEN 3 SEPARATE REGIONS
C     OF THE (X,NU) SPACE.
C     BRANCH TO THE PROPER COMPUTATIONAL METHOD
C     DEPENDING ON THE REGION.
C     NUCUT HAS THE VALUE 1000.
C
      IF(NU.LT.NUCUT)GOTO1000
      IF(NU.GE.NUCUT.AND.X.LE.ANU)GOTO2000
      IF(NU.GE.NUCUT.AND.X.GT.ANU)GOTO3000
      IBRAN=1
      WRITE(IPR,99)IBRAN
   99 FORMAT(1H ,42H*****INTERNAL ERROR IN CHSCDF SUBROUTINE--,
     146HIMPOSSIBLE BRANCH CONDITION AT BRANCH POINT = ,I8)
      RETURN
C
C     TREAT THE SMALL AND MODERATE DEGREES OF FREEDOM CASE
C     (THAT IS, WHEN NU IS SMALLER THAN 1000).
C     METHOD UTILIZED--EXACT FINITE SUM
C     (SEE AMS 55, PAGE 941, FORMULAE 26.4.4 AND 26.4.5).
C
 1000 CONTINUE
      CHI=DSQRT(DX)
      IEVODD=NU-2*(NU/2)
      IF(IEVODD.EQ.0)GOTO120
C
      SUM=0.0D0
      TERM=1.0/CHI
      IMIN=1
      IMAX=NU-1
      GOTO130
C
  120 SUM=1.0D0
      TERM=1.0D0
      IMIN=2
      IMAX=NU-2
C
  130 IF(IMIN.GT.IMAX)GOTO160
      DO100I=IMIN,IMAX,2
      AI=I
      TERM=TERM*(DX/AI)
      SUM=SUM+TERM
  100 CONTINUE
  160 CONTINUE
C
      SUM=SUM*DEXP(-DX/2.0D0)
      IF(IEVODD.EQ.0)GOTO170
      SUM=(DSQRT(2.0D0/PI))*SUM
```

```
      SPCHI=CHI
      CDFN=GAUCDF(SPCHI)
      DCDFN=CDFN
      SUM=SUM+2.0D0*(1.0D0-DCDFN)
  170 CHSCDF=1.0D0-SUM
      RETURN
C
C     TREAT THE CASE WHEN NU IS LARGE
C     (THAT IS, WHEN NU IS EQUAL TO OR GREATER THAN 1000)
C     AND X IS LESS THAN OR EQUAL TO NU.
C     METHOD UTILIZED--WILSON-HILFERTY APPROXIMATION
C     (SEE JOHNSON AND KOTZ, VOLUME 1, PAGE 176, FORMULA 28).
C
 2000 CONTINUE
      DFACT=4.5D0*DNU
      U=(((DX/DNU)**DPOWER)-1.0D0+(1.0D0/DFACT))*DSQRT(DFACT)
      CDFN=GAUCDF(U)
      CHSCDF=CDFN
      RETURN
C
C     TREAT THE CASE WHEN NU IS LARGE
C     (THAT IS, WHEN NU IS EQUAL TO OR GREATER THAN 1000)
C     AND X IS LARGER THAN NU.
C     METHOD UTILIZED--HILL'S ASYMPTOTIC EXPANSION
C     (SEE JOHNSON AND KOTZ, VOLUME 1, PAGE 180, FORMULA 33.1).
C
 3000 CONTINUE
      DW=DSQRT(DX-DNU-DNU*DLOG(DX/DNU))
      DANU=DSQRT(2.0D0/DNU)
      D1=DW
      D2=DW**2
      D3=DW**3
      TERM0=DW
      TERM1=B11*DANU
      TERM2=B21*D1*(DANU**2)
      TERM3=B31*(D2+B32)*(DANU**3)
      TERM4=B41*(B42*D3+B43*D1)*(DANU**4)
      U=TERM0+TERM1+TERM2+TERM3+TERM4
      CDFN=GAUCDF(U)
      CHSCDF=CDFN
      RETURN
C
      END
```

```
      REAL FUNCTION FCDF(X,NU1,NU2)
C
C     PURPOSE--THIS SUBROUTINE COMPUTES THE CUMULATIVE DISTRIBUTION
C              FUNCTION VALUE FOR THE F DISTRIBUTION
C              WITH INTEGER DEGREES OF FREEDOM
C              PARAMETERS = NU1 AND NU2.
C              THIS DISTRIBUTION IS DEFINED FOR ALL NON-NEGATIVE X.
C              THE PROBABILITY DENSITY FUNCTION IS GIVEN
C              IN THE REFERENCES BELOW.
C     INPUT  ARGUMENTS--X      = THE SINGLE PRECISION VALUE AT
C                                WHICH THE CUMULATIVE DISTRIBUTION
C                                FUNCTION IS TO BE EVALUATED.
C                                X SHOULD BE NON-NEGATIVE.
C                     --NU1    = THE INTEGER DEGREES OF FREEDOM
C                                FOR THE NUMERATOR OF THE F RATIO.
C                                NU1 SHOULD BE POSITIVE.
C                     --NU2    = THE INTEGER DEGREES OF FREEDOM
C                                FOR THE DENOMINATOR OF THE F RATIO.
C                                NU2 SHOULD BE POSITIVE.
C     OUTPUT--THE SINGLE PRECISION CUMULATIVE DISTRIBUTION
C             FUNCTION VALUE CDF FOR THE F DISTRIBUTION
C             WITH DEGREES OF FREEDOM
C             PARAMETERS = NU1 AND NU2.
C     PRINTING--NONE UNLESS AN INPUT ARGUMENT ERROR CONDITION EXISTS.
C     RESTRICTIONS--X SHOULD BE NON-NEGATIVE.
C                 --NU1 SHOULD BE A POSITIVE INTEGER VARIABLE.
C                 --NU2 SHOULD BE A POSITIVE INTEGER VARIABLE.
C     OTHER DATAPAC   SUBROUTINES NEEDED--GAUCDF,CHSCDF.
C     FORTRAN LIBRARY SUBROUTINES NEEDED--DSQRT, DATAN.
C     MODE OF INTERNAL OPERATIONS--DOUBLE PRECISION.
C     LANGUAGE--ANSI FORTRAN.
C     REFERENCES--NATIONAL BUREAU OF STANDARDS APPLIED MATHEMATICS
C                 SERIES 55, 1964, PAGES 946-947,
C                 FORMULAE 26.6.4, 26.6.5, 26.6.8, AND 26.6.15.
C               --JOHNSON AND KOTZ, CONTINUOUS UNIVARIATE
C                 DISTRIBUTIONS--2, 1970, PAGE 83, FORMULA 20,
C                 AND PAGE 84, THIRD FORMULA.
C               --PAULSON, AN APPROXIMATE NORMAILIZATION
C                 OF THE ANALYSIS OF VARIANCE DISTRIBUTION,
C                 ANNALS OF MATHEMATICAL STATISTICS, 1942,
C                 NUMBER 13, PAGES 233-135.
C               --SCHEFFE AND TUKEY, A FORMULA FOR SAMPLE SIZES
C                 FOR POPULATION TOLERANCE LIMITS, 1944,
C                 NUMBER 15, PAGE 217.
C     WRITTEN BY--JAMES J. FILLIBEN
C                 STATISTICAL ENGINEERING LABORATORY (205.03)
C                 NATIONAL BUREAU OF STANDARDS
C                 WASHINGTON, D. C. 20234
C                 PHONE:  301-921-2315
C     ORIGINAL VERSION--AUGUST    1972.
C     UPDATED         --SEPTEMBER 1975.
C     UPDATED         --NOVEMBER  1975.
C     UPDATED         --OCTOBER   1976.
C     CONVERTED TO REAL FUNCTION BY E. PRINCE, JULY 23, 1980.
C
C-----------------------------------------------------------------------
C
      DOUBLE PRECISION DX,PI,ANU1,ANU2,Z,SUM,TERM,AI,COEF1,COEF2,ARG
```

```
      DOUBLE PRECISION COEF
      DOUBLE PRECISION THETA, SINTH, COSTH, A, B
      DOUBLE PRECISION DSQRT, DATAN
      DOUBLE PRECISION DFACT1, DFACT2, DNUM, DDEN
      DOUBLE PRECISION DPOW1, DPOW2
      DOUBLE PRECISION DNU1, DNU2
      DOUBLE PRECISION TERM1, TERM2, TERM3
      DATA PI/3.14159265358979D0/
      DATA DPOW1, DPOW2/0.33333333333333D0, 0.66666666666667D0/
      DATA NUCUT1, NUCUT2/100, 1000/
      DATA IPR/6/
C
C     CHECK THE INPUT ARGUMENTS FOR ERRORS
C
      IF(NU1.LE.0)GOTO50
      IF(NU2.LE.0)GOTO55
      IF(X.LT.0.0)GOTO60
      GOTO90
   50 WRITE(IPR,15)
      WRITE(IPR,47)NU1
      CDF=0.0
      RETURN
   55 WRITE(IPR,23)
      WRITE(IPR,47)NU2
      CDF=0.0
      RETURN
   60 WRITE(IPR,4)
      WRITE(IPR,46)X
      CDF=0.0
      RETURN
   90 CONTINUE
    4 FORMAT(1H , 96H***** NON-FATAL DIAGNOSTIC--THE FIRST  INPUT ARGUME
     1NT TO THE FCDF   SUBROUTINE IS NEGATIVE *****)
   15 FORMAT(1H , 91H***** FATAL ERROR--THE SECOND INPUT ARGUMENT TO THE
     1 FCDF   SUBROUTINE IS NON-POSITIVE *****)
   23 FORMAT(1H , 91H***** FATAL ERROR--THE THIRD  INPUT ARGUMENT TO THE
     1 FCDF   SUBROUTINE IS NON-POSITIVE *****)
   46 FORMAT(1H , 35H***** THE VALUE OF THE ARGUMENT IS ,E15.8,6H *****)
   47 FORMAT(1H , 35H***** THE VALUE OF THE ARGUMENT IS ,I8   ,6H *****)
C
C-----START POINT-----------------------------------------------------
C
      DX=X
      M=NU1
      N=NU2
      ANU1=NU1
      ANU2=NU2
      DNU1=NU1
      DNU2=NU2
C
C     IF X IS NON-POSITIVE, SET CDF = 0.0 AND RETURN.
C     IF NU2 IS 5 THROUGH 9 AND X IS MORE THAN 3000
C     STANDARD DEVIATIONS BELOW THE MEAN,
C     SET CDF = 0.0 AND RETURN.
C     IF NU2 IS 10 OR LARGER AND X IS MORE THAN 150
C     STANDARD DEVIATIONS BELOW THE MEAN,
C     SET CDF = 0.0 AND RETURN.
C     IF NU2 IS 5 THROUGH 9 AND X IS MORE THAN 3000
C     STANDARD DEVIATIONS ABOVE THE MEAN,
```

```
C       SET CDF = 1.0 AND RETURN.
C       IF NU2 IS 10 OR LARGER AND X IS MORE THAN 150
C       STANDARD DEVIATIONS ABOVE THE MEAN,
C       SET CDF = 1.0 AND RETURN.
C
        IF(X.LE.0.0)GOTO105
        IF(NU2.LE.4)GOTO109
        T1=2.0/ANU1
        T2=ANU2/(ANU2-2.0)
        T3=(ANU1+ANU2-2.0)/(ANU2-4.0)
        AMEAN=T2
        SD=SQRT(T1*T2*T2*T3)
        ZRATIO=(X-AMEAN)/SD
        IF(NU2.LT.10.AND.ZRATIO.LT.-3000.0)GOTO105
        IF(NU2.GE.10.AND.ZRATIO.LT.-150.0)GOTO105
        IF(NU2.LT.10.AND.ZRATIO.GT.3000.0)GOTO107
        IF(NU2.GE.10.AND.ZRATIO.GT.150.0)GOTO107
        GOTO109
  105 FCDF=0.0
        RETURN
  107 FCDF=1.0
        RETURN
  109 CONTINUE
C
C       DISTINGUISH BETWEEN 6 SEPARATE REGIONS
C       OF THE (NU1,NU2) SPACE.
C       BRANCH TO THE PROPER COMPUTATIONAL METHOD
C       DEPENDING ON THE REGION.
C       NUCUT1 HAS THE VALUE 100.
C       NUCUT2 HAS THE VALUE 1000.
C
        IF(NU1.LT.NUCUT2.AND.NU2.LT.NUCUT2)GOTO1000
        IF(NU1.GE.NUCUT2.AND.NU2.GE.NUCUT2)GOTO2000
        IF(NU1.LT.NUCUT1.AND.NU2.GE.NUCUT2)GOTO3000
        IF(NU1.GE.NUCUT1.AND.NU2.GE.NUCUT2)GOTO2000
        IF(NU1.GE.NUCUT2.AND.NU2.LT.NUCUT1)GOTO5000
        IF(NU1.GE.NUCUT2.AND.NU2.GE.NUCUT1)GOTO2000
        IBRAN=5
        WRITE(IPR,99)IBRAN
   99 FORMAT(1H ,42H*****INTERNAL ERROR IN   FCDF SUBROUTINE--,
     146HIMPOSSIBLE BRANCH CONDITION AT BRANCH POINT = ,I8)
        RETURN
C
C       TREAT THE CASE WHEN NU1 AND NU2
C       ARE BOTH SMALL OR MODERATE
C       (THAT IS, BOTH ARE SMALLER THAN 1000).
C       METHOD UTILIZED--EXACT FINITE SUM
C       (SEE AMS 55, PAGE 946, FORMULAE 26.6.4, 26.6.5,
C       AND 26.6.8).
C
 1000 CONTINUE
        Z=ANU2/(ANU2+ANU1*DX)
        IFLAG1=NU1-2*(NU1/2)
        IFLAG2=NU2-2*(NU2/2)
        IF(IFLAG1.EQ.0)GOTO120
        IF(IFLAG2.EQ.0)GOTO150
        GOTO250
C
C       DO THE NU1 EVEN AND NU2 EVEN OR ODD CASE
```

```
C
  120 SUM=0.0D0
      TERM=1.0D0
      IMAX=(M-2)/2
      IF(IMAX.LE.0)GOTO110
      DO100I=1,IMAX
      AI=I
      COEF1=2.0D0*(AI-1.0D0)
      COEF2=2.0D0*AI
      TERM=TERM*((ANU2+COEF1)/COEF2)*(1.0D0-Z)
      SUM=SUM+TERM
  100 CONTINUE
C
  110 SUM=SUM+1.0D0
      SUM=(Z**(ANU2/2.0D0))*SUM
      FCDF=1.0D0-SUM
      RETURN
C
C     DO THE NU1 ODD AND NU2 EVEN CASE
C
  150 SUM=0.0D0
      TERM=1.0D0
      IMAX=(N-2)/2
      IF(IMAX.LE.0)GOTO210
      DO200I=1,IMAX
      AI=I
      COEF1=2.0D0*(AI-1.0D0)
      COEF2=2.0D0*AI
      TERM=TERM*((ANU1+COEF1)/COEF2)*Z
      SUM=SUM+TERM
  200 CONTINUE
C
  210 SUM=SUM+1.0D0
      FCDF=((1.0D0-Z)**(ANU1/2.0D0))*SUM
      RETURN
C
C     DO THE NU1 ODD AND NU2 ODD CASE
C
  250 SUM=0.0D0
      TERM=1.0D0
      ARG=DSQRT((ANU1/ANU2)*DX)
      THETA=DATAN(ARG)
      SINTH=ARG/DSQRT(1.0D0+ARG*ARG)
      COSTH=1.0D0/DSQRT(1.0D0+ARG*ARG)
      IF(N.EQ.1)GOTO320
      IF(N.EQ.3)GOTO310
      IMAX=N-2
      DO300I=3,IMAX,2
      AI=I
      COEF1=AI-1.0D0
      COEF2=AI
      TERM=TERM*(COEF1/COEF2)*(COSTH*COSTH)
      SUM=SUM+TERM
  300 CONTINUE
C
  310 SUM=SUM+1.0D0
      SUM=SUM*SINTH*COSTH
C
  320 A=(2.0D0/PI)*(THETA+SUM)
```

```
  350 SUM=0.0D0
      TERM=1.0D0
      IF(M.EQ.1)B=0.0D0
      IF(M.EQ.1)GOTO450
      IF(M.EQ.3)GOTO410
      IMAX=M-3
      DO400I=1,IMAX,2
      AI=I
      COEF1=AI
      COEF2=AI+2.0D0
      TERM=TERM*((ANU2+COEF1)/COEF2)*(SINTH*SINTH)
      SUM=SUM+TERM
  400 CONTINUE
C
  410 SUM=SUM+1.0D0
      SUM=SUM*SINTH*(COSTH**N)
      COEF=1.0D0
      IEVODD=N-2*(N/2)
      IMIN=3
      IF(IEVODD.EQ.0)IMIN=2
      IF(IMIN.GT.N)GOTO420
      DO430I=IMIN,N,2
      AI=I
      COEF=((AI-1.0D0)/AI)*COEF
  430 CONTINUE
C
  420 COEF=COEF*ANU2
      IF(IEVODD.EQ.0)GOTO440
      COEF=COEF*(2.0D0/PI)
C
  440 B=COEF*SUM
C
  450 FCDF=A-B
      RETURN
C
C     TREAT THE CASE WHEN NU1 AND NU2
C     ARE BOTH LARGE
C     (THAT IS, BOTH ARE EQUAL TO OR LARGER THAN 1000);
C     OR WHEN NU1 IS MODERATE AND NU2 IS LARGE
C     (THAT IS, WHEN NU1 IS EQUAL TO OR GREATER THAN 100
C     BUT SMALLER THAN 1000,
C     AND NU2 IS EQUAL TO OR LARGER THAN 1000);
C     OR WHEN NU2 IS MODERATE AND NU1 IS LARGE
C     (THAT IS WHEN NU2 IS EQUAL TO OR GREATER THAN 100
C     BUT SMALLER THAN 1000,
C     AND NU1 IS EQUAL TO OR LARGER THAN 1000).
C     METHOD UTILIZED--PAULSON APPROXIMATION
C     (SEE AMS 55, PAGE 947, FORMULA 26.6.15).
C
 2000 CONTINUE
      DFACT1=1.0D0/(4.5D0*DNU1)
      DFACT2=1.0D0/(4.5D0*DNU2)
      DNUM=((1.0D0-DFACT2)*(DX**DPOW1))-(1.0D0-DFACT1)
      DDEN=DSQRT((DFACT2*(DX**DPOW2))+DFACT1)
      U=DNUM/DDEN
      GCDF=GAUCDF(U)
      FCDF=GCDF
      RETURN
C
```

```
C       TREAT THE CASE WHEN NU1 IS SMALL
C       AND NU2 IS LARGE
C       (THAT IS, WHEN NU1 IS SMALLER THAN 100,
C       AND NU2 IS EQUAL TO OR LARGER THAN 1000).
C       METHOD UTILIZED--SHEFFE-TUKEY APPROXIMATION
C       (SEE JOHNSON AND KOTZ, VOLUME 2, PAGE 84, THIRD FORMULA).
C
 3000 CONTINUE
      TERM1=DNU1
      TERM2=(DNU1/DNU2)*(0.5D0*DNU1-1.0D0)
      TERM3=-(DNU1/DNU2)*0.5D0
      U=(TERM1+TERM2)/((1.0D0/DX)-TERM3)
      CCDF=CHSCDF(U,NU1)
      FCDF=CCDF
      RETURN
C
C       TREAT THE CASE WHEN NU2 IS SMALL
C       AND NU1 IS LARGE
C       (THAT IS, WHEN NU2 IS SMALLER THAN 100,
C       AND NU1 IS EQUAL TO OR LARGER THAN 1000).
C       METHOD UTILIZED--SHEFFE-TUKEY APPROXIMATION
C       (SEE JOHNSON AND KOTZ, VOLUME 2, PAGE 84, THIRD FORMULA).
C
 5000 CONTINUE
      TERM1=DNU2
      TERM2=(DNU2/DNU1)*(0.5D0*DNU2-1.0D0)
      TERM3=-(DNU2/DNU1)*0.5D0
      U=(TERM1+TERM2)/(DX-TERM3)
      CCDF=CHSCDF(U,NU2)
      FCDF=1.0-CCDF
      RETURN
C
      END
```

```
      REAL FUNCTION GAUCDF(X)
C
C     PURPOSE--THIS SUBROUTINE COMPUTES THE CUMULATIVE DISTRIBUTION
C              FUNCTION VALUE FOR THE NORMAL (GAUSSIAN)
C              DISTRIBUTION WITH MEAN = 0 AND STANDARD DEVIATION = 1.
C              THIS DISTRIBUTION IS DEFINED FOR ALL X AND HAS
C              THE PROBABILITY DENSITY FUNCTION
C              F(X) = (1/SQRT(2*PI))*EXP(-X*X/2).
C     INPUT  ARGUMENTS--X      = THE SINGLE PRECISION VALUE AT
C                                WHICH THE CUMULATIVE DISTRIBUTION
C                                FUNCTION IS TO BE EVALUATED.
C     OUTPUT--THE SINGLE PRECISION CUMULATIVE DISTRIBUTION
C             FUNCTION VALUE CDF.
C     PRINTING--NONE.
C     RESTRICTIONS--NONE.
C     OTHER DATAPAC   SUBROUTINES NEEDED--NONE.
C     FORTRAN LIBRARY SUBROUTINES NEEDED--EXP.
C     MODE OF INTERNAL OPERATIONS--SINGLE PRECISION.
C     LANGUAGE--ANSI FORTRAN.
C     REFERENCES--NATIONAL BUREAU OF STANDARDS APPLIED MATHEMATICS
C                 SERIES 55, 1964, PAGE 932, FORMULA 26.2.17.
C               --JOHNSON AND KOTZ, CONTINUOUS UNIVARIATE
C                 DISTRIBUTIONS--1, 1970, PAGES 40-111.
C     WRITTEN BY--JAMES J. FILLIBEN
C                 STATISTICAL ENGINEERING LABORATORY (205.03)
C                 NATIONAL BUREAU OF STANDARDS
C                 WASHINGTON, D. C. 20234
C                 PHONE:  301-921-2315
C     ORIGINAL VERSION--JUNE      1972.
C     UPDATED         --SEPTEMBER 1975.
C     UPDATED         --NOVEMBER  1975.
C     CONVERTED TO REAL FUNCTION BY E. PRINCE, JULY 18, 1980.
C
C---------------------------------------------------------------------
C
      DATA B1,B2,B3,B4,B5,P/.319381530,-0.356563782,1.781477937,-1.82125
     15978,1.330274429,.2316419/
      DATA IPR/6/
C
C     CHECK THE INPUT ARGUMENTS FOR ERRORS.
C     NO INPUT ARGUMENT ERRORS POSSIBLE
C     FOR THIS DISTRIBUTION.
C
C-----START POINT-----------------------------------------------------
C
      Z=X
      IF(X.LT.0.0)Z=-Z
      T=1.0/(1.0+P*Z)
      CDF=1.0-((0.39894228040143  )*EXP(-0.5*Z*Z))*(B1*T+B2*T**2+B3*
     1T**3+B4*T**4+B5*T**5)
      IF(X.LT.0.0)CDF=1.0-CDF
      GAUCDF=CDF
C
      RETURN
      END
```

```
      REAL FUNCTION TCDF(X,NU)
C
C     PURPOSE--THIS SUBROUTINE COMPUTES THE CUMULATIVE DISTRIBUTION
C              FUNCTION VALUE FOR STUDENT'S T DISTRIBUTION
C              WITH INTEGER DEGREES OF FREEDOM PARAMETER = NU.
C              THIS DISTRIBUTION IS DEFINED FOR ALL X.
C              THE PROBABILITY DENSITY FUNCTION IS GIVEN
C              IN THE REFERENCES BELOW.
C     INPUT  ARGUMENTS--X      = THE SINGLE PRECISION VALUE AT
C                                WHICH THE CUMULATIVE DISTRIBUTION
C                                FUNCTION IS TO BE EVALUATED.
C                                X SHOULD BE NON-NEGATIVE.
C                     --NU     = THE INTEGER NUMBER OF DEGREES
C                                OF FREEDOM.
C                                NU SHOULD BE POSITIVE.
C     OUTPUT--THE SINGLE PRECISION CUMULATIVE DISTRIBUTION
C             FUNCTION VALUE CDF FOR THE STUDENT'S T DISTRIBUTION
C             WITH DEGREES OF FREEDOM PARAMETER = NU.
C     PRINTING--NONE UNLESS AN INPUT ARGUMENT ERROR CONDITION EXISTS.
C     RESTRICTIONS--NU SHOULD BE A POSITIVE INTEGER VARIABLE.
C     OTHER DATAPAC   SUBROUTINES NEEDED--GAUCDF.
C     FORTRAN LIBRARY SUBROUTINES NEEDED--DSQRT, DATAN.
C     MODE OF INTERNAL OPERATIONS--DOUBLE PRECISION.
C     LANGUAGE--ANSI FORTRAN.
C     REFERENCES--NATIONAL BUREAU OF STANDARDS APPLIED MATHMATICS
C                 SERIES 55, 1964, PAGE 948, FORMULAE 26.7.3 AND 26.7.4.
C               --JOHNSON AND KOTZ, CONTINUOUS UNIVARIATE
C                 DISTRIBUTIONS--2, 1970, PAGES 94-129.
C               --FEDERIGHI, EXTENDED TABLES OF THE
C                 PERCENTAGE POINTS OF STUDENT'S
C                 T-DISTRIBUTION, JOURNAL OF THE
C                 AMERICAN STATISTICAL ASSOCIATION,
C                 1959, PAGES 683-688.
C               --OWEN, HANDBOOK OF STATISTICAL TABLES,
C                 1962, PAGES 27-30.
C               --PEARSON AND HARTLEY, BIOMETRIKA TABLES
C                 FOR STATISTICIANS, VOLUME 1, 1954,
C                 PAGES 132-134.
C     WRITTEN BY--JAMES J. FILLIBEN
C                 STATISTICAL ENGINEERING LABORATORY (205.03)
C                 NATIONAL BUREAU OF STANDARDS
C                 WASHINGTON, D. C. 20234
C                 PHONE:  301-921-2315
C     ORIGINAL VERSION--JUNE      1972.
C     UPDATED         --MAY       1974.
C     UPDATED         --SEPTEMBER 1975.
C     UPDATED         --NOVEMBER  1975.
C     UPDATED         --OCTOBER   1976.
C     CONVERTED TO REAL FUNCTION BY E. PRINCE, JULY 23, 1980.
C
C-----------------------------------------------------------------------
C
      DOUBLE PRECISION DX,DNU,PI,C,CSQ,S,SUM,TERM,AI
      DOUBLE PRECISION DSQRT,DATAN
      DOUBLE PRECISION DCONST
      DOUBLE PRECISION TERM1,TERM2,TERM3
      DOUBLE PRECISION DCDFN
      DOUBLE PRECISION DCDF
```

```
      DOUBLE PRECISION B11
      DOUBLE PRECISION B21,B22,B23,B24,B25
      DOUBLE PRECISION B31,B32,B33,B34,B35,B36,B37
      DOUBLE PRECISION D1,D3,D5,D7,D9,D11
      DATA NUCUT/1000/
      DATA PI/3.14159265358979D0/
      DATA DCONST/0.3989422804D0/
      DATA B11/0.25D0/
      DATA B21/0.01041666666667D0/
      DATA B22,B23,B24,B25/3.0D0,-7.0D0,-5.0D0,-3.0D0/
      DATA B31/0.00260416666667D0/
      DATA B32,B33,B34,B35,B36,B37/1.0D0,-11.0D0,14.0D0,6.0D0,
     1                             -3.0D0,-15.0D0/
      DATA IPR/6/
C
C     CHECK THE INPUT ARGUMENTS FOR ERRORS
C
      IF(NU.LE.0)GOTO50
      GOTO90
   50 WRITE(IPR,15)
      WRITE(IPR,47)NU
      TCDF=0.0
      RETURN
   90 CONTINUE
   15 FORMAT(1H , 91H***** FATAL ERROR--THE SECOND INPUT ARGUMENT TO THE
     1 TCDF    SUBROUTINE IS NON-POSITIVE *****)
   47 FORMAT(1H , 35H***** THE VALUE OF THE ARGUMENT IS ,I8   ,6H *****)
C
C-----START POINT-----------------------------------------------------
C
      DX=X
      ANU=NU
      DNU=NU
C
C     IF NU IS 3 THROUGH 9 AND X IS MORE THAN 3000
C     STANDARD DEVIATIONS BELOW THE MEAN,
C     SET CDF = 0.0 AND RETURN.
C     IF NU IS 10 OR LARGER AND X IS MORE THAN 150
C     STANDARD DEVIATIONS BELOW THE MEAN,
C     SET CDF = 0.0 AND RETURN.
C     IF NU IS 3 THROUGH 9 AND X IS MORE THAN 3000
C     STANDARD DEVIATIONS ABOVE THE MEAN,
C     SET CDF = 1.0 AND RETURN.
C     IF NU IS 10 OR LARGER AND X IS MORE THAN 150
C     STANDARD DEVIATIONS ABOVE THE MEAN,
C     SET CDF = 1.0 AND RETURN.
C
      IF(NU.LE.2)GOTO109
      SD=SQRT(ANU/(ANU-2.0))
      Z=X/SD
      IF(NU.LT.10.AND.Z.LT.-3000.0)GOTO107
      IF(NU.GE.10.AND.Z.LT.-150.0)GOTO107
      IF(NU.LT.10.AND.Z.GT.3000.0)GOTO108
      IF(NU.GE.10.AND.Z.GT.150.0)GOTO108
      GOTO109
  107 TCDF=0.0
      RETURN
  108 TCDF=1.0
      RETURN
```

```
  109 CONTINUE
C
C     DISTINGUISH BETWEEN THE SMALL AND MODERATE
C     DEGREES OF FREEDOM CASE VERSUS THE
C     LARGE DEGREES OF FREEDOM CASE
C
      IF(NU.LT.NUCUT)GOTO110
      GOTO250
C
C     TREAT THE SMALL AND MODERATE DEGREES OF FREEDOM CASE
C     METHOD UTILIZED--EXACT FINITE SUM
C     (SEE AMS 55, PAGE 948, FORMULAE 26.7.3 AND 26.7.4).
C
  110 CONTINUE
      C=DSQRT(DNU/(DX*DX+DNU))
      CSQ=DNU/(DX*DX+DNU)
      S=DX/DSQRT(DX*DX+DNU)
      IMAX=NU-2
      IEVODD=NU-2*(NU/2)
      IF(IEVODD.EQ.0)GOTO120
C
      SUM=C
      IF(NU.EQ.1)SUM=0.0D0
      TERM=C
      IMIN=3
      GOTO130
C
  120 SUM=1.0D0
      TERM=1.0D0
      IMIN=2
C
  130 IF(IMIN.GT.IMAX)GOTO160
      DO100I=IMIN,IMAX,2
      AI=I
      TERM=TERM*((AI-1.0D0)/AI)*CSQ
      SUM=SUM+TERM
  100 CONTINUE
C
  160 SUM=SUM*S
      IF(IEVODD.EQ.0)GOTO170
      SUM=(2.0D0/PI)*(DATAN(DX/DSQRT(DNU))+SUM)
  170 TCDF=0.5D0+SUM/2.0D0
      RETURN
C
C     TREAT THE LARGE DEGREES OF FREEDOM CASE.
C     METHOD UTILIZED--TRUNCATED ASYMPTOTIC EXPANSION
C     (SEE JOHNSON AND KOTZ, VOLUME 2, PAGE 102, FORMULA 10;
C     SEE FEDERIGHI, PAGE 687).
C
  250 CONTINUE
      CDFN=GAUCDF(X)
      DCDFN=CDFN
      D1=DX
      D3=DX**3
      D5=DX**5
      D7=DX**7
      D9=DX**9
      D11=DX**11
      TERM1=B11*(D3+D1)/DNU
```

```
      TERM2=B21*(B22*D7+B23*D5+B24*D3+B25*D1)/(DNU**2)
      TERM3=B31*(B32*D11+B33*D9+B34*D7+B35*D5+B36*D3+B37*D1)/(DNU**3)
      DCDF=TERM1+TERM2+TERM3
      DCDF=DCDFN-(DCONST*(DEXP(-DX*DX/2.0D0)))*DCDF
      TCDF=DCDF
      RETURN
C
      END
```

```
      SUBROUTINE CORLIB(R3,ELT,RC)
C     THIS SUBROUTINE APPLIES THE LIBRATION CORRECTION TO AN ATOM
C     WHOSE POSITION IS DEFINED, WITH RESPECT TO THE ORIGIN OF A
C     RIGID BODY, BY THE VECTOR RC, WHICH IS A THREE DIMENSIONAL
C     VECTOR WITH UNITS OF LENGTH. THE TRUE, EQUILIBRIUM POSITION OF
C     THE ATOM IS RETURNED IN THE THREE DIMENSIONAL VECTOR R3. EL IS
C     THE LIBRATION TENSOR. IT IS A 3X3 MATRIX, AND HAS UNITS OF
C     RADIANS**2.
      DIMENSION R3(3,3),ELT(3,3),RC(3),ACON(3,3),BCON(3)
      DO 30 I=1,3
      BCON(I)=RC(I)
      DO 30 J=1,3
30    ACON(I,J)=0.
40    DO 50 I=1,3
      DO 50 J=1,3
50    ACON(I,J)=ACON(I,J)+EIJ(ELT,I,J)
      CALL SOLVE(ACON,BCON,R3)
      RETURN
      END

      FUNCTION EIJ(EL,I,J)
C     THIS FUNCTION SUPPLIES THE ELEMENTS OF THE MATRIX RELATING THE
C     EQUILIBRIUM POSITION OF AN ATOM IN A LIBRATING, RIGID BODY TO
C     ITS TIME-AVERAGED POSITION.
      DIMENSION EL(3,3)
      IF(I.EQ.J)GO TO 20
      EIJ=.5*EL(I,J)-(1./24.)*(EL4(EL,I,1,1,J)+EL4(EL,I,2,2,J)+EL4(EL,
     1 I,3,3,J))
      RETURN
20    K=MOD(I,3)+1
      L=MOD(I+1,3)+1
      EIJ=1.-.5*(EL(K,K)+EL(L,L))+(1./24.)*(EL4(EL,I,I,K,K)+EL4(EL,I,I,
     1 L,L)+3.*(EL(K,K)**2+EL(L,L)**2)+2.*EL4(EL,K,K,L,L))
      RETURN
      END

      FUNCTION EL4(EL,I,J,K,L)
      DIMENSION EL(3,3)
      EL4=EL(I,J)*EL(K,L)+EL(I,K)*EL(J,L)+EL(I,L)*EL(J,K)
      RETURN
      END

      SUBROUTINE SOLVE(A,B,C)
      DIMENSION A(3,3),B(3),C(3)
      DELTA=DETERM(A(1,1),A(1,2),A(1,3))
      C(1)=DETERM(B(1),A(1,2),A(1,3))/DELTA
      C(2)=DETERM(A(1,1),B(1),A(1,3))/DELTA
      C(3)=DETERM(A(1,1),A(1,2),B(1))/DELTA
      RETURN
      END
```

```
      FUNCTION DETERM(C1,C2,C3)
      DIMENSION C1(3),C2(3),C3(3)
      DETERM=C1(1)*C2(2)*C3(3)+C1(2)*C2(3)*C3(1)+C1(3)*C2(1)*C3(2)
     1 -C1(3)*C2(2)*C3(1)-C1(2)*C2(1)*C3(3)-C1(1)*C2(3)*C3(2)
      RETURN
      END

      FUNCTION SKTLS(R,I,J,T,EL,S)
C     THIS FUNCTION GIVES THE IJ ELEMENT OF THE SECOND CUMULANT TENSOR
C     FOR AN ATOM AT POSITION R IN A RIGID BODY DUE TO THE RIGID BODY
C     MOTIONS DESCRIBED BY TENSORS T, L, AND S.  ALL VECTOR AND TENSOR
C     ELEMENTS ASSUME AN ORTHONORMAL COORDINATE SYSTEM, AND ANGULAR
C     DISPLACEMENTS ARE IN RADIANS.
      DIMENSION R(3),T(3,3),EL(3,3),S(3,3),RR(3)
      B(I1,I2,I3)=BIJK(RR,I1,I2,I3)
      C(I1,I2,I3,I4)=CIJKL(RR,I1,I2,I3,I4)
      DO 5 M=1,3
    5 RR(M)=R(M)
      SUM=T(I,J)
      DO 10 K=1,3
      SUM=SUM+AIJ(R,J,K)*S(K,I)+AIJ(R,I,K)*S(K,J)
      DO 10 L=1,3
      SUM=SUM+AIJ(R,I,K)*AIJ(R,J,L)*EL(K,L)
      DO 10 M=1,3
      SUM=SUM+3.*EL(L,M)*(S(K,I)*C(J,K,L,M)+S(K,J)*C(I,K,L,M))
      DO 10 N=1,3
   10 SUM=SUM+EL(K,L)*EL(M,N)*(3.*(AIJ(R,I,K)*C(J,L,M,N)+AIJ(R,J,K)
     1    *C(I,L,M,N))+2.*B(I,K,M)*B(J,L,N))
      SKTLS=SUM
      RETURN
      END

      FUNCTION D2KDT(I,J,K,L)
C     THE DERIVATIVE OF SECOND CUMULANT TENSOR ELEMENT IJ WITH RESPECT
C     TO T(K,L).
      D2KDT=0.
      IF(I.EQ.K.AND.J.EQ.L)D2KDT=1.
      RETURN
      END
```

```
      FUNCTION D2KDL(R,EL,S,I,J,K,L)
C     THE DERIVATIVE OF SECOND CUMULANT TENSOR ELEMENT IJ WITH RESPECT
C     TO L(K,L).
      DIMENSION R(3),EL(3,3),S(3,3),RR(3)
      C(I,J,K,L)=CIJKL(RR,I,J,K,L)
      C6(I,K,L,M)=6.*C(I,K,L,M)
      ACC(I,J,K,L,M,N)=3.*AIJ(RR,I,K)*C(J,L,M,N)
      DO 10 M=1,3
   10 RR(M)=R(M)
      SUM=AIJ(R,I,K)*AIJ(R,J,L)+AIJ(R,I,L)*AIJ(R,J,K)
      DO 1000 M=1,3
      SUM=SUM+S(M,I)*C6(J,K,L,M)+S(M,J)*C6(I,K,L,M)
      DO 1000 N=1,3
      SUM=SUM+EL(M,N)*(ACC(I,J,K,L,M,N)+ACC(I,J,L,K,M,N)+ACC(J,I,K,L,M,N
     1      )+ACC(J,I,L,K,M,N)+ACC(I,J,M,N,K,L)+ACC(J,I,M,N,K,L)+ACC(I,J
     2      ,M,N,L,K)+ACC(J,I,M,N,L,K)+4.*(BIJK(R,I,K,M)*BIJK(R,J,L,N)+
     3      BIJK(R,I,L,M)*BIJK(R,J,K,N)))
 1000 CONTINUE
      D2KDL=SUM
      IF(K.EQ.L)D2KDL=SUM/2.
      RETURN
      END

      FUNCTION D2KDS(R,EL,I,J,K,L)
C     THE DERIVATIVE OF SECOND CUMULANT TENSOR ELEMENT IJ WITH RESPECT
C     TO S(K,L).
      DIMENSION R(3),EL(3,3),RR(3)
      CCC(J,K,L,M)=3.*CIJKL(RR,J,K,L,M)
      DO 10 M=1,3
   10 RR(M)=R(M)
      SUM=0.
      IF(I.NE.L)GO TO 1000
      SUM=SUM+AIJ(R,J,K)
      DO 500 M=1,3
      DO 500 N=1,3
  500 SUM=SUM+EL(M,N)*CCC(J,K,M,N)
 1000 IF(J.NE.L)GO TO 2000
      SUM=SUM+AIJ(R,I,K)
      DO 1500 M=1,3
      DO 1500 N=1,3
 1500 SUM=SUM+EL(M,N)*CCC(I,K,M,N)
 2000 D2KDS=SUM
      RETURN
      END
```

```
      FUNCTION TKTLS(R, I, J, K, EL, S)
C     THIS FUNCTION GIVES THE IJK ELEMENT OF THE THIRD CUMULANT TENSOR
C     FOR AN ATOM AT POSITION R IN A RIGID BODY DUE TO THE RIGID BODY
C     MOTION TENSORS, L AND S.
      DIMENSION R(3), EL(3, 3), S(3, 3)
      INTEGER P
      SUM=0.
      DO 1000 L=1, 3
      DO 1000 M=1, 3
      SUM=SUM+2. *(S(L, I)*S(M, J)*BIJK(R, K, L, M)
     1           +S(L, I)*S(M, K)*BIJK(R, J, L, M)
     2           +S(L, J)*S(M, K)*BIJK(R, I, L, M))
      DO 1000 N=1, 3
      SUM=SUM+2. *EL(M, N)*
     1    (S(L, I)*(AIJ(R, J, M)*BIJK(R, K, L, N)+AIJ(R, K, M)*BIJK(R, J, L, N))
     2    +S(L, J)*(AIJ(R, I, M)*BIJK(R, K, L, N)+AIJ(R, K, M)*BIJK(R, I, L, N))
     3    +S(L, K)*(AIJ(R, I, M)*BIJK(R, J, L, N)+AIJ(R, J, M)*BIJK(R, I, L, N)))
      DO 1000 P=1, 3
      SUM=SUM+EL(N, P)*(S(L, J)*S(M, K)*PIJKLM(R, I, L, M, N, P)+S(L, I)*S(M, K)*
     1    PIJKLM(R, J, L, M, N, P)+S(L, I)*S(M, J)*PIJKLM(R, K, L, M, N, P)
     2    +2. *EL(L, M)*(AIJ(R, I, L)*AIJ(R, J, N)*BIJK(R, K, M, P)
     3    +AIJ(R, I, L)*AIJ(R, K, N)*BIJK(R, J, M, P)
     4    +AIJ(R, J, L)*AIJ(R, K, N)*BIJK(R, I, M, P)))
1000  CONTINUE
      TKTLS=SUM
      RETURN
      END

      FUNCTION D3KDL(R, EL, S, I, J, K, L, M)
C     THE DERIVATIVE OF THIRD CUMULANT TENSOR ELEMENT IJK WITH RESPECT
C     TO L(L, M).
      DIMENSION R(3), EL(3, 3), S(3, 3), RR(3)
      INTEGER P
      ABB(J, K, L, M, N)=2. *AIJ(RR, J, L)*BIJK(RR, K, M, N)
      AAB(I, J, K, L, M, N, NN)=AIJ(RR, I, L)*AIJ(RR, J, N)*BIJK(RR, K, M, NN)
      PP(I, J, K, L, M)=PIJKLM(RR, I, J, K, L, M)+PIJKLM(RR, I, J, K, M, L)
      DO 10 N=1, 3
   10 RR(N)=R(N)
      SUM=0.
      DO 1000 N=1, 3
      SUM=SUM+S(N, I)*(ABB(J, K, L, M, N)+ABB(K, J, L, M, N)+ABB(J, K, M, L, N)+ABB(K
     1 , J, M, L, N))+S(N, J)*(ABB(I, K, L, M, N)+ABB(K, I, L, M, N)+ABB(I, K, M, L, N
     2    )+ABB(K, I, M, L, N))+S(N, K)*(ABB(I, J, L, M, N)+ABB(J, I, L, M, N)+ABB(I,
     3    J, M, L, N)+ABB(J, I, M, L, N))
      DO 1000 P=1, 3
      SUM=SUM+S(N, J)*S(P, K)*PP(I, N, P, L, M)+S(N, I)*S(P, K)*PP(J, N, P, L, M)
     1    +S(N, I)*S(P, J)*PP(K, N, P, L, M)+2. *EL(N, P)*(AAB(I, J, K, L, M, N, P)
     2    +AAB(I, K, J, L, M, N, P)+AAB(J, K, I, L, M, N, P)+AAB(I, J, K, N, P, L, M)
     3    +AAB(I, K, J, N, P, L, M)+AAB(J, K, I, N, P, L, M))
1000  CONTINUE
      D3KDL=SUM
      IF(L. EQ. M)D3KDL=SUM/2.
      RETURN
      END
```

```
      FUNCTION D3KDS(R,EL,S,I,J,K,LP,MP)
C     THE DERIVATIVE OF THIRD CUMULANT TENSOR ELEMENT IJK WITH RESPECT
C     TO S(LP,MP).
      DIMENSION R(3),EL(3,3),S(3,3),RR(3)
      INTEGER P
      BB(K,L,M)=2.*BIJK(RR,K,L,M)
      ABB(J,K,L,M,N)=2.*AIJ(RR,J,L)*BIJK(RR,K,M,N)
      DO 10 M=1,3
   10 RR(M)=R(M)
      SUM=0.
      DO 1000 L=1,3
      DO 1000 M=1,3
      IF(L.EQ.LP.AND.I.EQ.MP)SUM=SUM+S(M,J)*BB(K,L,M)+S(M,K)*BB(J,L,M)
      IF(M.EQ.LP.AND.J.EQ.MP)SUM=SUM+S(L,I)*BB(K,L,M)
      IF(M.EQ.LP.AND.K.EQ.MP)SUM=SUM+S(L,I)*BB(J,L,M)+S(L,J)*BB(I,L,M)
      IF(L.EQ.LP.AND.J.EQ.MP)SUM=SUM+S(M,K)*BB(I,L,M)
      DO 1000 N=1,3
      IF(L.NE.LP)GO TO 500
      IF(I.EQ.MP)SUM=SUM+EL(M,N)*(ABB(J,K,M,L,N)+ABB(K,J,M,L,N))
      IF(J.EQ.MP)SUM=SUM+EL(M,N)*(ABB(I,K,M,L,N)+ABB(K,I,M,L,N))
      IF(K.EQ.MP)SUM=SUM+EL(M,N)*(ABB(I,J,M,L,N)+ABB(J,I,M,L,N))
500   DO 1000 P=1,3
      IF(L.NE.LP)GO TO 750
      IF(J.EQ.MP)SUM=SUM+EL(N,P)*S(M,K)*PIJKLM(R,I,L,M,N,P)
      IF(I.EQ.MP)SUM=SUM+EL(N,P)*(S(M,K)*PIJKLM(R,J,L,M,N,P)+S(M,J)*
     1            PIJKLM(R,K,L,M,N,P))
750   IF(M.NE.LP)GO TO 1000
      IF(K.EQ.MP)SUM=SUM+EL(N,P)*(S(L,J)*PIJKLM(R,I,L,M,N,P)+S(L,I)*
     1            PIJKLM(R,J,L,M,N,P))
      IF(J.EQ.MP)SUM=SUM+EL(N,P)*S(L,I)*PIJKLM(R,K,L,M,N,P)
1000  CONTINUE
      D3KDS=SUM
      RETURN
      END
```

```
      SUBROUTINE EULER(PHI,CHI,OMEGA,TRANS)
C     SUBROUTINE TO GENERATE THE MATRIX B-INVERSE FROM THE EULERIAN
C     ANGLES PHI,CHI,AND OMEGA. OMEGA IS THE ANGLE (IN DEGREES) BY
C     WHICH A SPECIAL, ORTHONORMAL COORDINATE SYSTEM MUST BE ROTATED,
C     CLOCKWISE AS VIEWED DOWN THE POSITIVE Z AXIS OF A 'STANDARD',
C     REFERENCE COORDINATE SYSTEM, IN ORDER TO BRING THE Z AXIS OF THE
C     SPECIAL SYSTEM INTO THE X - Z PLANE OF THE STANDARD SYSTEM. CHI
C     IS THE ANGLE (IN DEGREES) BY WHICH THE SPECIAL SYSTEM MUST BE
C     ROTATED CLOCKWISE ABOUT THE Y AXIS OF THE STANDARD SYSTEM TO
C     BRING THE Z AXES OF THE TWO COORDINATE SYSTEMS INTO COINCIDENCE.
C     PHI IS THE ANGLE (IN DEGREES) OF A CLOCKWISE ROTATION ABOUT THE
C     COMMON Z AXES TO BRING THE TWO COORDINATE SYSTEMS INTO COIN-
C     CIDENCE. ENTRIES DTDPHI, DTDCHI, AND DTDOMG GIVE THE MATRICES
C     WHOSE ELEMENTS ARE THE DERIVATIVES OF THE EULER MATRIX WITH
C     RESPECT TO PHI, CHI, AND OMEGA.
      DATA RAD/57.2957792/
      DIMENSION TRANS(3,3)
      ASSIGN 10 TO KGO
      GO TO 100
10    TRANS(1,1)=CP*CX*CO-SP*SO
      TRANS(2,1)=CP*CX*SO+SP*CO
      TRANS(3,1)=-CP*SX
      TRANS(1,2)=-SP*CX*CO-CP*SO
      TRANS(2,2)=-SP*CX*SO+CP*CO
      TRANS(3,2)=SP*SX
      TRANS(1,3)=SX*CO
      TRANS(2,3)=SX*SO
      TRANS(3,3)=CX
      RETURN
      ENTRY DTDPHI(PHI,CHI,OMEGA,TRANS)
      ASSIGN 20 TO KGO
      GO TO 100
20    TRANS(1,1)=(-SP*CX*CO-CP*SO)/RAD
      TRANS(2,1)=(-SP*CX*SO+CP*CO)/RAD
      TRANS(3,1)=SP*SX/RAD
      TRANS(1,2)=(-CP*CX*CO+SP*SO)/RAD
      TRANS(2,2)=(-CP*CX*SO-SP*CO)/RAD
      TRANS(3,2)=CP*SX/RAD
      DO 25 I=1,3
25    TRANS(I,3)=0.
      RETURN
      ENTRY DTDCHI(PHI,CHI,OMEGA,TRANS)
      ASSIGN 30 TO KGO
      GO TO 100
30    TRANS(1,1)=-CP*SX*CO/RAD
      TRANS(2,1)=-CP*SX*SO/RAD
      TRANS(3,1)=-CP*CX/RAD
      TRANS(1,2)=SP*SX*CO/RAD
      TRANS(2,2)=SP*SX*SO/RAD
      TRANS(3,2)=SP*CX/RAD
      TRANS(1,3)=CX*CO/RAD
      TRANS(2,3)=CX*SO/RAD
      TRANS(3,3)=-SX/RAD
      RETURN
      ENTRY DTDOMG(PHI,CHI,OMEGA,TRANS)
      ASSIGN 40 TO KGO
      GO TO 100
40    TRANS(1,1)=(-CP*CX*SO-SP*CO)/RAD
```

```
      TRANS(2,1)=(CP*CX*CO-SP*SO)/RAD
      TRANS(1,2)=(SP*CX*SO-CP*CO)/RAD
      TRANS(2,2)=(-SP*CX*CO-CP*SO)/RAD
      TRANS(1,3)=-SX*SO/RAD
      TRANS(2,3)=SX*CO/RAD
      DO 45 I=1,3
45    TRANS(3,I)=0.
      RETURN
100   CP=COS(PHI/RAD)
      CX=COS(CHI/RAD)
      CO=COS(OMEGA/RAD)
      SP=SIN(PHI/RAD)
      SX=SIN(CHI/RAD)
      SO=SIN(OMEGA/RAD)
      GO TO KGO(10,20,30,40)
      END

      FUNCTION PIJKLM(R,I,J,K,L,M)
      DIMENSION R(3)
      PIJKLM=12.*DIJKLM(R,I,J,K,L,M)
      RETURN
      END
```

```
      FUNCTION AIJ(R, I, J)
      DIMENSION R(3)
      GO TO (10, 20, 30), I
10    GO TO (100, 200, 300), J
20    GO TO (300, 100, 200), J
30    GO TO (200, 300, 100), J
100   AIJ=0.
      RETURN
200   MM=MOD(I+1, 3)+1
      AIJ=R(MM)
      RETURN
300   MM=MOD(I, 3)+1
      AIJ=-R(MM)
      RETURN
      END

      FUNCTION BIJK(R, I, JJ, KK)
      DIMENSION R(3)
      DIV=2.
      IF(JJ.EQ.KK)DIV=1.
      J=MIN0(JJ, KK)
      K=MAX0(JJ, KK)
      GO TO (10, 20, 30), I
10    GO TO (12, 14, 400), J
12    GO TO (100, 200, 300), K
14    IF(K-2)400, 400, 100
20    GO TO (22, 24, 400), J
22    GO TO (400, 300, 100), K
24    IF(K-2)100, 100, 200
30    GO TO (32, 34, 100), J
32    GO TO (400, 100, 200), K
34    IF(K-2)400, 400, 300
100   BIJK=0.
      RETURN
200   MM=MOD(I, 3)+1
      BIJK=.5*R(MM)/DIV
      RETURN
300   MM=MOD(I+1, 3)+1
      BIJK=.5*R(MM)/DIV
      RETURN
  400 BIJK=-.5*R(I)/DIV
      RETURN
      END
```

```
      FUNCTION CIJKL(R, I, JJ, KK, LL)
      DIMENSION R(3)
      IF(JJ.EQ.KK.AND.JJ.EQ.LL)GO TO 7
      IF(JJ.NE.KK.AND.JJ.NE.LL.AND.KK.NE.LL)GO TO 100
      IF(JJ-KK)1,2,3
    1 J=JJ
      L=KK
      K=LL
      GO TO 4
    2 J=MINO(JJ,LL)
      K=KK
      L=MAXO(JJ,LL)
      GO TO 4
    3 J=KK
      K=LL
      L=JJ
    4 DIV=3.
      GO TO 9
    7 J=JJ
      K=JJ
      L=JJ
      DIV=1.
9     GO TO (10,20,30),I
10    GO TO (12,15,300),J
12    GOTO (13,100,100),K
13    GO TO (100,200,300),L
15    IF(K-2)16,16,200
16    IF(L-2)200,200,300
20    GO TO (22,25,200),J
22    GOTO (23,24,300),K
23    GO TO (300,100,200),L
24    IF(L-2)300,300,100
25    IF(K-2)26,26,100
26    IF(L-2)100,100,200
30    GO TO (32,35,100),J
32    GO TO (33,34,200),K
33    GO TO (200,300,100),L
34    IF(L-2)200,200,100
35    IF(K-2)36,36,300
36    IF(L-2)300,300,100
100   CIJKL=0.
      RETURN
200   MM=MOD(I+1,3)+1
      CIJKL=-R(MM)/(6.*DIV)
      RETURN
300   MM=MOD(I,3)+1
      CIJKL=R(MM)/(6.*DIV)
      RETURN
      END
```

```
      FUNCTION DIJKLM(R,I,JI,KI,LI,MI)
      DIMENSION R(3)
      IN=JI*KI*LI*MI
      IF(MOD(IN,6).EQ.0.AND.IN.LE.18)GO TO 1080
      IF(IN.EQ.4.OR.IN.EQ.9.OR.IN.EQ.36)GO TO 1060
      IF(JI.NE.KI.OR.KI.NE.LI.OR.LI.NE.MI)GO TO 1040
      DIV=1.
      GO TO 1100
 1040 DIV=4.
      GO TO 1100
 1060 DIV=6.
      GO TO 1100
 1080 DIV=12.
 1100 IF(MOD(IN,3).NE.0)GO TO 1500
      M=3
      IN=IN/3
      IF(MOD(IN,3).NE.0)GO TO 1250
      L=3
      IN=IN/3
      IF(MOD(IN,3).NE.0)GO TO 1125
      K=3
      J=IN/3
      GO TO 1900
 1125 IF(MOD(IN,2).NE.0)GO TO 1150
      K=2
      J=IN/2
      GO TO 1900
 1150 K=1
      J=1
      GO TO 1900
 1250 IF(MOD(IN,2).NE.0)GO TO 1375
      L=2
      IN=IN/2
      IF(MOD(IN,2).NE.0)GO TO 1150
      K=2
      J=IN/2
      GO TO 1900
 1375 L=1
      GO TO 1150
 1500 IF(MOD(IN,2).NE.0)GO TO 1750
      M=2
      IN=IN/2
      GO TO 1250
 1750 M=1
      GO TO 1375
 1900 GO TO (100,200,300),I
 100  GO TO (110,150,2000),J
 110  GO TO (120,140,3000),K
 120  GO TO (125,130,2000),L
 125  GO TO (1000,4000,3000),M
 130  IF(M-2)1000,2000,1000
 140  IF(L-2)1000,145,4000
 145  IF(M-2)1000,4000,3000
 150  IF(K-2)1000,160,1000
 160  IF(L-2)1000,170,5000
 170  IF(M-2)1000,2000,1000
 200  GO TO (210,250,2000),J
 210  GO TO (220,240,1000),K
```

```
220    GO TO (225,230,5000),L
225    GO TO (2000,3000,1000),M
230    IF(M-2)1000,2000,4000
240    IF(L-2)1000,245,3000
245    IF(M-2)1000,3000,1000
250    IF(K-2)1000,260,4000
260    IF(L-2)1000,270,2000
270    IF(M-2)1000,1000,4000
300    GO TO (310,350,1000),J
310    GO TO (320,340,4000),K
320    GO TO (325,330,2000),L
325    GO TO (2000,1000,4000),M
330    IF(M-2)1000,5000,3000
340    IF(L-2)1000,345,1000
345    IF(M-2)1000,1000,4000
350    IF(K-2)1000,360,3000
360    IF(L-2)1000,370,2000
370    IF(M-2)1000,2000,3000
1000   DIJKLM=0.
       RETURN
2000   FAC=1.
 2010  DIJKLM=FAC*R(I)/(24.*DIV)
       RETURN
3000   MM=MOD(I+1,3)+1
 3010  DIJKLM=-R(MM)/(24.*DIV)
       RETURN
4000   MM=MOD(I,3)+1
       GO TO 3010
5000   FAC=2.
       GO TO 2010
       END
```

Bibliography

The following list makes no pretense to being inclusive. Some books are current, and some are out of date. Some are classics in their fields, while others are obscure. All contain at least a few pages of useful information.

Ahmed, F. R. (ed.): Crystallographic Computing. Proceedings of the 1969 International Summer School on Crystallographic Computing. Munksgaard, Copenhagen, 1970.

Arndt, U. W., and Willis, B. T. M.: Single Crystal Diffractometry. Cambridge University Press, Cambridge, 1966.

Draper, N., and Smith, H.: Applied Regression Analysis. John Wiley & Sons, New York, London, Sydney, 1966.

Eisele, J. A., and Mason, R. M.: Applied Matrix and Tensor Analysis. Wiley-Interscience, New York, London, Sydney, Toronto, 1970.

Hamilton, W. C.: Statistics in Physical Science. The Ronald Press Company, New York, 1964.

International Tables for X-ray Crystallography, published by Kynoch Press, Birmingham, for the International Union of Crystallography.

Kendall, M. G., and Stuart, A.: The Advanced Theory of Statistics, 2nd ed. Charles Griffin & Co., Ltd., London, 1963.

Lide, D. R., Jr., and Paul, M. A. (eds.): Critical Evaluation of Chemical and Physical Structural Information. National Academy of Sciences, Washington, 1974.

Murray, W. (ed.): Numerical Methods for Unconstrained Optimization. Academic Press, London, New York, 1972.

Nye, J. F.: Physical Properties of Crystals. Clarendon Press, Oxford, 1957.

Pauling, L., and Wilson, E. B. Jr.: Introduction to Quantum Mechanics. McGraw-Hill Book Company, New York, London, 1935.

Phillips, F. C.: An Introduction to Crystallography. Longmans, Green & Co., London, New York, Toronto, 1946.

Pipes, L. A.: Applied Mathematics for Engineers and Physicists. McGraw-Hill Book Company, New York, London, 1946.

Squires, G. L.: Introduction to the Theory of Thermal Neutron Scattering. Cambridge University Press, Cambridge, London, New York, Melbourne, 1978.

Stewart, G. W.: Introduction to Matrix Computations. Academic Press, New York, London, 1973.

Widder, D. V.: Advanced Calculus. Prentice-Hall, New York, 1947.

Willis, B. T. M. (ed.): Thermal Neutron Diffraction. Oxford University Press, Oxford, 1970.

Wilson, E. B., Jr., Decius, J. C. and Cross, P. C.: Molecular Vibrations, The Theory of Infrared and Raman Vibrational Spectra. McGraw-Hill Book Company, New York, Toronto, London, 1955.

Wooster, W. A.: Crystal Physics. Cambridge University Press, Cambridge, 1949.

Index